Mt. Wanda Historical Ecology Investigation

A reconnaissance study investigating historical landscape data for the John Muir National Historic Site

PREPARED FOR THE NATIONAL PARK SERVICE

MARCH 2015

SFEI
A⬡S⬡C
SAN FRANCISCO ESTUARY INSTITUTE
AQUATIC SCIENCE CENTER

Prepared by:

Sean Baumgarten

Erin Beller

Robin Grossinger

DESIGN AND PRODUCTION • RUTH ASKEVOLD

National Park Service
U.S. Department of Interior

San Francisco Estuary Institute Publication #743

SUGGESTED CITATION

Baumgarten SA, EE Beller, RM Grossinger, RA Askevold. 2015. Mt. Wanda Historical Ecology Investigation. A Report of SFEI-ASC's Resilient Landscapes Program, Publication #743, San Francisco Estuary Institute-Aquatic Science Center, Richmond, CA.

REPORT AVAILABILITY

Report is available on SFEI's website at http://www.sfei.org/MtWanda.

COVER CREDITS

Front cover, left to right, top to bottom: ca. 1890 photo of the Muir House and the northern slope of Mt. Wanda (image courtesy of John Muir NHS, JOMU 1729); *Dodecatheon hendersonii* (photo by Tom Hilton); John Muir near his home in Martinez, CA, circa 1907-9 (photo #F24-1307, John Muir Papers, Holt-Atherton Special Collections, University of the Pacific Library. ©1984 Muir-Hanna Trust); 1933 Wieslander VTM map (courtesy of Information Center for the Environment, UC Davis); 1891 map of Alhambra Ranch (Brown 1891, image courtesy of John Muir NHS, JOMU 3889); Strentzel-Muir Ranch, 1885 (HABS CAL,7-MART,2--1, courtesy of Library of Congress, Prints and Photographs Division); topography of Mt. Wanda; USGS [1893-4]1897 topographic map overlaid on 2014 imagery (Google Earth 2014); oaks and grassland on Mt. Wanda (photo by SFEI, July 2013).

Back cover: Portion of John Muir's journal entry from January 24, 1895, describing view from Mt. Wanda (Muir 1895, John Muir Papers, Holt-Atherton Special Collections, University of the Pacific Library. ©1984 Muir-Hanna Trust).

Pages ii-iii: 1891 map of Alhambra Ranch (Brown 1891, image courtesy of John Muir NHS, JOMU 3889).

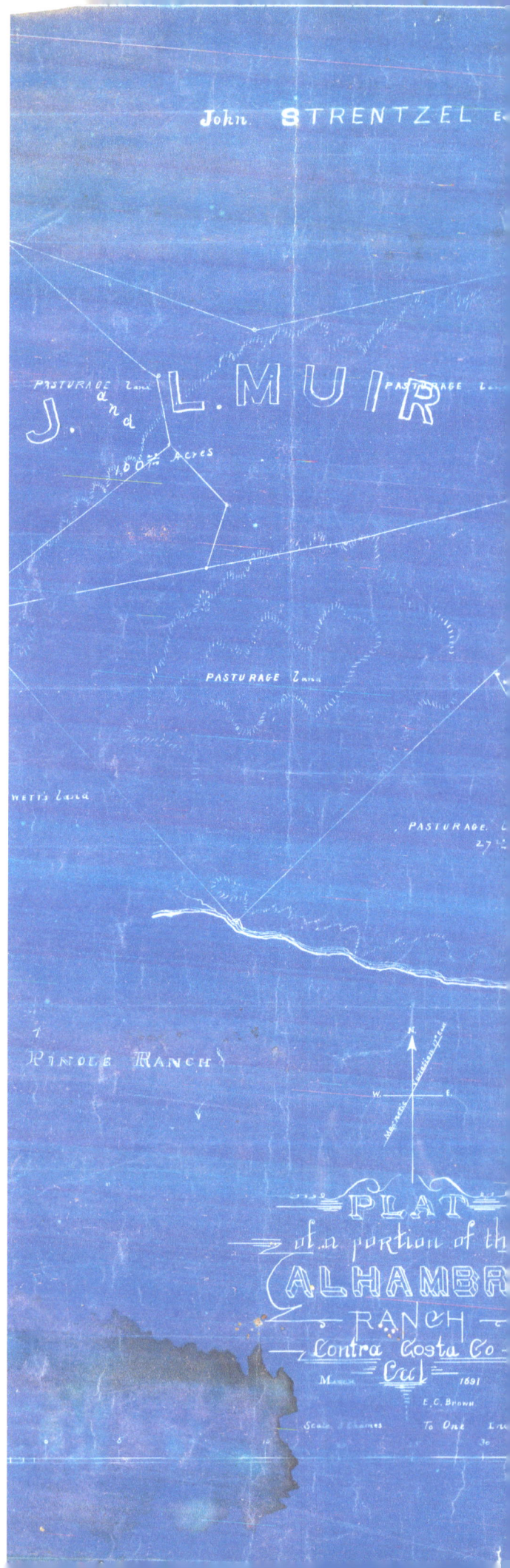

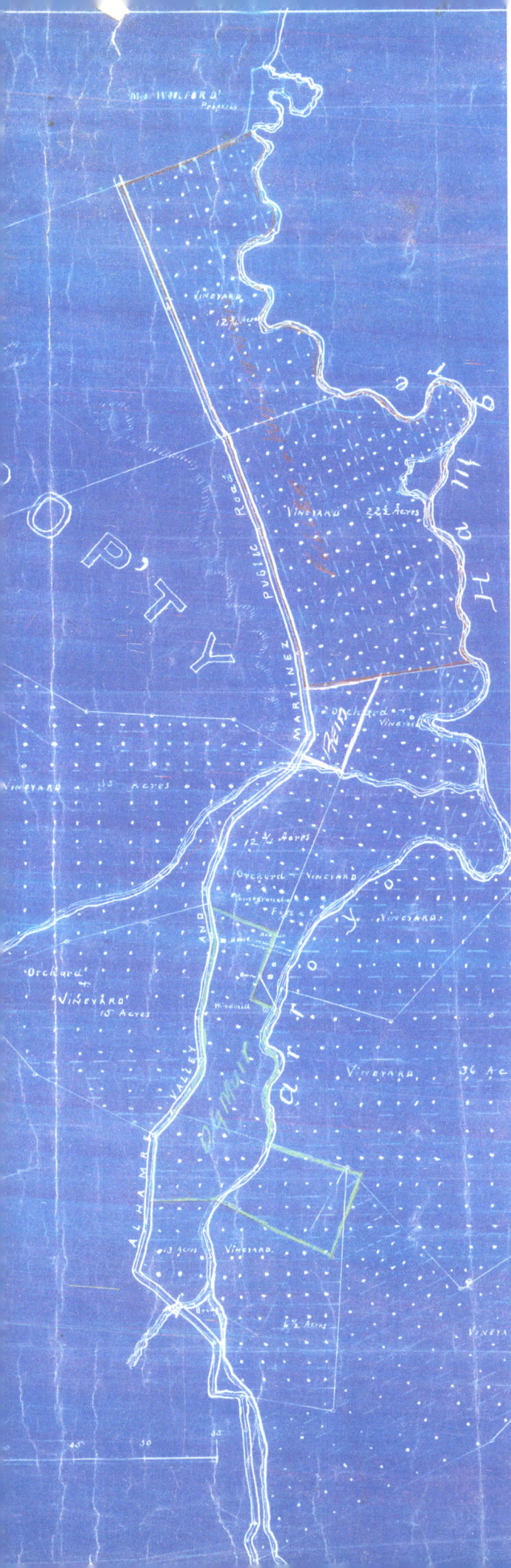

CONTENTS

INTRODUCTION 1

   Overview 1

   Study Area Description 2

   Land Use History 4

METHODS 6

ASSESSMENT OF HISTORICAL DATASET 8

   Maps 8

   Photographs 16

   Texts 21

PRELIMINARY DATA SYNTHESIS 32

NEXT STEPS..38

ACKNOWLEDGMENTS 40

REFERENCES..41

APPENDIX: SELECTED QUOTES 44

John Muir leaning on an oak tree in Martinez, CA, circa 1905. (Photo #F23-1287, John Muir Papers, Holt-Atherton Special Collections, University of the Pacific Library. ©1984 Muir-Hanna Trust)

INTRODUCTION

Overview

On April 12, 1895, the renowned naturalist John Muir wrote in his journal, "Took a fine fragrant walk up the West hills with Wanda + Helen who I am glad to see love walking – flowers – trees + every bird + beast + creeping thing" (Muir 1895). The journal entry describes one of the many walks that Muir took with his daughters Wanda and Helen on the hills behind their home in the Alhambra Valley, at the southern end of Martinez, California. Nearly 120 years after he wrote these words, visitors to the John Muir National Historic Site can literally follow in John Muir's footsteps, strolling along the network of trails that wind up the slopes of Mt. Wanda, which Muir named after his eldest daughter. With its stately oak woodlands, diverse wildflower displays, and expansive views of Mt. Diablo and east Contra Costa County, Mt. Wanda is a valuable remnant of California's past landscape and offers a unique opportunity to explore the relatively undisturbed backyard of one of America's most iconic environmental figures.

The National Park Service (NPS) manages the John Muir National Historic Site (JOMU), including the 326-acre Mt. Wanda unit, to approximate the conditions that existed from the time of Spanish settlement through the time that John Muir lived on the ranch (1849-1914). To achieve this objective and to support resource management efforts, NPS requires detailed information about the ecological and physical characteristics of the historical landscape. Historical data can shed light on a range of questions relevant to site management: How has the distribution or composition of vegetation communities changed? Are there species that were documented historically that have been extirpated from the site? How have land use changes affected the physical and ecological environment on Mt. Wanda?

The goals of this reconnaissance study are to collect and assess historical data about the Mt. Wanda landscape, conduct an initial synthesis of the data to enhance understanding of early ecological conditions, and evaluate the potential benefits of further data collection and analysis. This report is intended to provide a concise summary of the collected data and preliminary findings, inform future ecological research and management activities, and serve as an educational and interpretive resource for JOMU visitors.

The report is organized as follows: the **Introduction** defines the study area and provides a brief summary of land use history. The **Methods** section explains the data collection, compilation, and synthesis process and lists source institutions visited or contacted. The **Assessment of Historical Dataset** section describes key sources and findings for each of the three major types of data collected: maps, photographs, and texts. **Preliminary Data Synthesis** summarizes the major findings of the study and examines how the Mt. Wanda landscape has changed over time. Finally, **Next Steps** outlines recommended steps for further research. The **Appendix** provides a compilation of key textual quotes.

Study Area Description

The John Muir National Historic Site is located in Contra Costa County to the south of downtown Martinez, CA. The site preserves the home and property where John Muir lived with his family from 1890 until his death in 1914. JOMU consists of three non-contiguous units: the John Muir House, just north of Highway 4; Mt. Wanda, just south of Highway 4; and the John Muir Gravesite, where John Muir and some of his family are buried, in Alhambra Valley just east of Mt. Wanda.

The Mt. Wanda parcel encompasses 326 acres, ranging in elevation from 120 to 670 ft (Fig. 1). Major vegetation communities on Mt. Wanda include blue oak woodland, valley oak woodland, coast live oak forest, California laurel forest, and grasslands (see Fig. 24 on page 37). Mt. Wanda lies within the 16.5 square mile Alhambra Creek watershed. Alhambra Creek borders the study area to the east, while Franklin Creek borders the study area on the north (both creeks are excluded from the study area). Numerous intermittent or ephemeral creeks flow off of Mt. Wanda (Moore 2006), the largest of which is Strentzel Creek, located in Strentzel Canyon on the southern side of the study area.

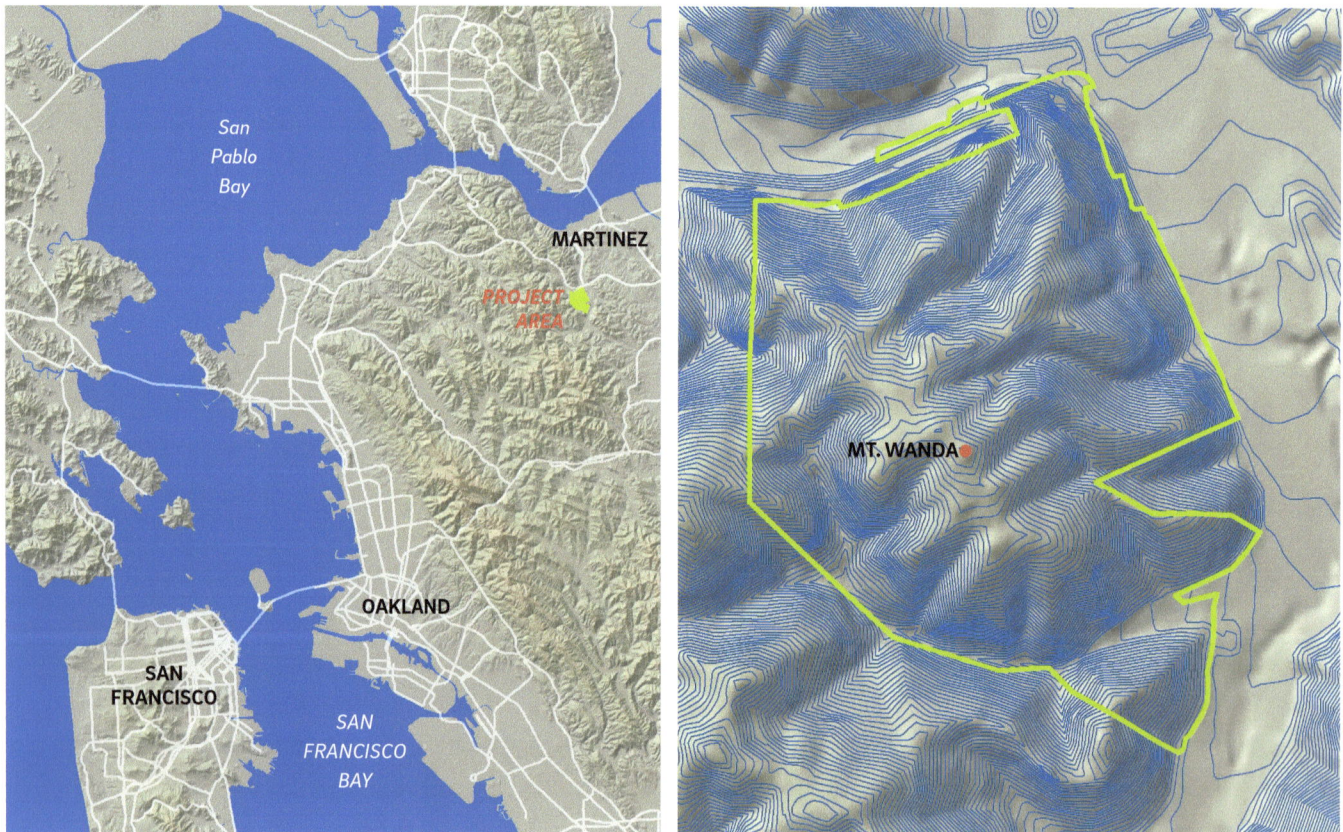

Figure 1. Study area. (above) Locator map and topography of Mt. Wanda study area. **(right)** The study area (outlined in green) encompasses 326 acres in Contra Costa County, and is the largest of the three units in the John Muir National Historic Site. (USDA 2009)

MARTINEZ

John Muir Pkwy.

Franklin Canyon Rd.

Franklin Creek

rlington Northern and Santa Fe Railway

Alhambra Ave.

Alhambra Creek

John Muir
National
Historic Site

MOUNT HELEN ▲

▲ MOUNT WANDA

Strentzel Creek

Alhambra Creek

N

1,000 feet
1:10,000 scale

Land Use History

Although Mt. Wanda has remained undeveloped, the landscape has nonetheless been impacted by a variety of land uses over the centuries. Prior to Spanish colonization, the Martinez area was occupied for many years by the native Karkin Ohlone (Milliken 1995, Killion 2005). Though there is no indication that major settlements existed in the Mt. Wanda area, the Ohlone likely influenced the landscape through practices such as burning, hunting, and acorn collection (Moore 2006).

Land use in the region changed dramatically with the establishment of Spanish missions and Mexican land grants in the late 18th and early 19th centuries, during which time livestock grazing became widespread. Mt. Wanda lies at the eastern edge of the Rancho El Pinole land grant, which was granted to Don Ignacio Martinez in 1823 (Killion 2005). Martinez used the rancho, including Mt. Wanda, to graze cattle, sheep, and other livestock. In 1837, for instance, Martinez stated that he had on Rancho El Pinole "more than three thousand head of cattle; four hundred head of horses, and six hundred head of sheep, and about eighty tame horses, together with more than three hundred head of milk cattle." It appears that the heaviest grazing may have occurred on portions of the Rancho in or near Alhambra Valley: Martinez stated that "the best part lies in the direction of the 'Sisca' and the 'Cañada del Hambre,' which is most frequented by the stock" (Martinez 1837).[1] Woodcutting and hay harvesting may also have occurred on Mt. Wanda in the mid-19th century (Hunter et al. 1993, O'Neil and Egan 2004).

Grazing likely continued on Mt. Wanda under various owners after the Martinez family sold the parcel in 1853 to Edward Franklin. In the mid-1880s, John Muir and his father-in-law, Dr. John Strentzel (Fig. 2), acquired the parcel and excluded livestock from Mt. Wanda in order to preserve the land in a more natural state (Killion 2005). Consequently, during the following decades Mt. Wanda remained ungrazed and mostly uncultivated, with the exception of a pear orchard on the lower slopes of the northern side and small vineyards and orchards on the southeastern side (Brown 1891, NPS 1991, Killion 2005). The other major land use change that occurred during this time was the construction in 1899 of a railroad trestle across Alhambra Valley and a railroad tunnel through the northernmost portion of Mt. Wanda (Keibel 1999, Killion 2005).

From the early 1900s until 1996, cattle again periodically grazed on Mt. Wanda under various landowners. Several remnant features associated with grazing during this period can still be seen on the landscape, including tanks, sheds, and cattle ponds (Killion 2005). In addition to periodic grazing, some areas of grassland near the summit of Mt. Wanda may have been cleared and graded for hay production during the 1920s (Moore 2006). On the southeastern side of Mt. Wanda, several structures were constructed on the Strain ranch property after 1930. Three natural gas wells were dug on Mt. Wanda in the early to mid-20th century, but never became operational (NPS 1991, Killion 2005). Mt. Wanda became part of the John Muir National Historic Site in 1991-2.

---

[1] "Sisca" refers to Franklin Canyon.

Figure 2. Dr. John Strentzel on the Strentzel-Muir Ranch, with the Martinez Adobe in the background, 1885. (HABS CAL,7-MART,2--1, courtesy of Library of Congress, Prints & Photographs Division)

Land Use Timeline

pre-1772	Martinez area occupied by Karkin Ohlone.
1772-6	Expeditions led by Spanish explorers Pedro Fages and Juan Bautista de Anza travel through the Martinez area.
1823	Rancho El Pinole granted to Ignacio Martinez.
1849	Vicente Martinez inherits a portion of the Rancho El Pinole, including Mt. Wanda; Martinez Adobe constructed.
1853	John and Louisiana Strentzel settle in Alhambra Valley.
1876	City of Martinez incorporated.
1880	John Muir marries Louie Strentzel and moves into the Alhambra ranch house (in Alhambra Valley just east of Mt. Wanda).
1882	Redfern Place (the structure that became the Muir House) constructed just north of Mt. Wanda.
1899	Alhambra Valley railroad trestle completed (rebuilt in 1929).
1964	John Muir National Historic Site designated. The site includes the Muir House, Martinez Adobe, and surrounding farmland.
1991-2	Mt. Wanda parcel added to the John Muir National Historic Site.

Sources: de Anza and Bolton 1930, Fages and Treutlein 1972, NPS 1991, Milliken 1995, Keibel 1999, Killion 2005

METHODS

The collection, compilation, and synthesis of historical data formed the major components of our research for this reconnaissance study. We assembled a wide range of historical and contemporary data, including maps (land grant case maps, General Land Office plats, U.S. Geological Survey [USGS] topographic maps, U.S. Department of Agriculture [USDA] soil surveys, Wieslander Vegetation Type Mapping [VTM], city and county maps, Sanborn maps, and property maps), photographs (19th-20th century landscape photos, 1939 aerial photos, and 2009 aerial photos), and texts (Spanish explorer accounts, land grant court case documents, land survey field notes, journal entries, letters, newspaper articles, species records, books, and technical reports) (Fig. 3). Data were collected from nine source institutions, either via electronic transfer or by taking digital photographs, obtaining scans, or transcribing text (Table 1). We also searched and/or collected data from numerous electronic databases, including Google Books, California Digital Newspaper Collection, Online Archive of California, David Rumsey Map Collection, USGS Bay Area Regional Database, and Consortium of California Herbaria. Data collection was not comprehensive: we focused on visiting the highest priority archives and collecting the most relevant sources, but a substantial amount of relevant data undoubtedly remains to be collected.

Once collected, data were sorted by source type, topic, and/or relevancy. Each source was reviewed and any pertinent information about the historical landscape or land uses was flagged. Relevant textual information was transcribed into a centralized textual data document, and spatially-locatable textual data were added to a GIS layer. We georeferenced seven high-priority maps in ArcGIS, and adjusted Wieslander VTM vectors to match georeferenced VTM rasters. Research for this study also drew heavily on county-scale data previously collected and compiled for the East Contra Costa County Historical Ecology Study (Stanford et al. 2011) and associated studies, including georeferenced maps, species records, and a 1939 aerial photomosaic.

Finally, the data were analyzed and cross-checked to document features of the historical landscape and identify changes in the landscape over time. Because this was a reconnaissance study using a limited dataset, we did not attempt to classify or map historical habitat types.

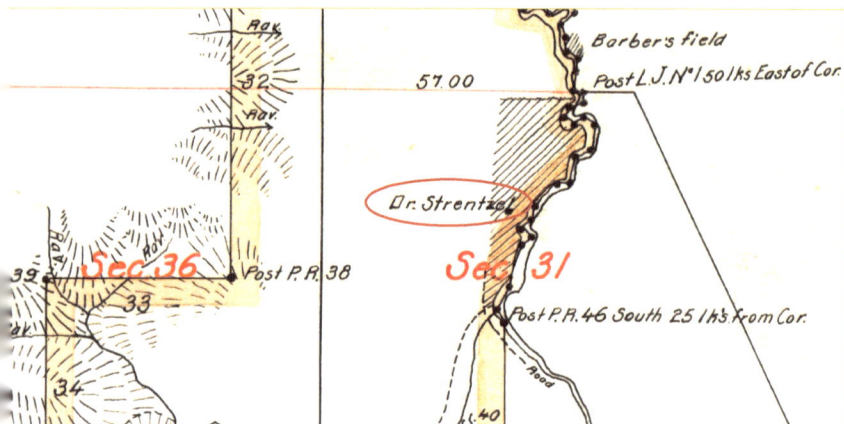

Source Institution	Location	Summary of Collected Materials
The Bancroft Library, UC Berkeley	Berkeley	Land grant case maps; county, municipal, and land grant survey maps; insurance maps; biographies; correspondence; land grant court case files and dockets; 19th-20th century landscape photographs
Bureau of Land Management	Sacramento	General Land Office field notes and survey plats
Contra Costa County Historical Society	Martinez	Regional histories; 19th-20th century landscape photographs
Contra Costa County Library	Multiple branches	20th century regional histories, reports, and biographies
Earth Sciences & Map Library, UC Berkeley	Berkeley	Insurance maps; geology maps; county maps
Holt-Atherton Special Collections, University of the Pacific Library	Stockton	Newspaper clippings; property survey notes and maps; deeds; 19th-20th century landscape photographs
Information Center for the Environment (ICE), UC Davis	Davis	Wieslander VTM raster and vector data
John Muir National Historic Site	Martinez	Technical reports; diaries; survey plats; property maps; county maps; lithographs; 19th-20th century landscape photographs; GIS layers
Shields Library, UC Davis	Davis	JOMU interpretive material; dissertations
Water Resources Collections and Archives	Riverside	Technical reports; hydrologic data; 20th century landscape photographs

Table 1. Source institutions visited/contacted for this study, and summary of collected materials. In addition to these materials, we reviewed data from numerous other source institutions that was previously collected and compiled as part of the East Contra Costa Historical Ecology Study (Stanford et al. 2011) and associated studies.

Figure 3. Historical and contemporary views of the Mt. Wanda area. (far left) Arroyo del Hambre and the residence of Dr. Strentzel are some of the features depicted on this 1865 plat of Rancho El Pinole. **(left)** Wanda and Helen Muir are shown in this late 1800s photograph, taken near the Martinez Adobe. The Muir House is visible in the background. **(right and far right)** Oak woodlands on the eastern slope of Mt. Wanda, July 2013. (Taylor 1865, courtesy of Bureau of Land Management; photo #F13-647, John Muir Papers, Holt-Atherton Special Collections, University of the Pacific Library. ©1984 Muir-Hanna Trust; photographs by SFEI, July 2013)

ASSESSMENT OF HISTORICAL DATASET

The following sections provide an overview of the historical data and discuss relevant ecological information about the early Mt. Wanda landscape. The discussion is organized around the three main types of historical data collected: maps, photographs, and texts (Fig. 4).

Maps

We collected approximately 60 maps for this study, ranging from land grant case maps to early county maps to federal surveys. In addition, we reviewed hundreds of maps (mostly parcel/property maps) previously collected as part of the East Contra Costa County Historical Ecology Study. Spanning a period of more than 170 years (ca. 1840 to present), the maps were produced for many disparate purposes and vary widely in terms of scale, accuracy, and level of detail. Table 2 summarizes the types of maps collected and reviewed for this study, organized chronologically.

Compared with landscape photos, aerial photographs, and textual data, historical maps provided relatively little information about the past landscape of Mt. Wanda. Unfortunately, the earliest maps of the area (land grant case maps, General Land Office plats) contained no relevant ecological information. Several characteristics of the Mt. Wanda parcel explain the relative paucity of ecological information available in historical maps, including its small size, upland topography (early maps are often more numerous and detailed for lowland areas than for less intensively developed upland areas), lack of major streams, and distance from major cities. Nevertheless, several of the collected maps, discussed below, do provide some clues about the character of the early landscape.

Source	Date Range	Source Institution(s)
Land grant case maps	ca. 1840-66	The Bancroft Library, UC Berkeley
General Land Office plats	1864-83	Bureau of Land Management; The Bancroft Library, UC Berkeley
County maps	1871-1940	The Bancroft Library, UC Berkeley; Contra Costa County Historical Society; Contra Costa County Library; Contra Costa County Public Works Department; Earth Sciences & Map Library, UC Berkeley; John Muir National Historic Site; UC Davis; Water Resources Collections and Archives
City and property maps	1873-1938	The Bancroft Library, UC Berkeley; Contra Costa County Clerk-Recorder; Contra Costa County Historical Society; John Muir National Historic Site
Sanborn maps	1891-7	The Bancroft Library, UC Berkeley; Earth Sciences & Map Library, UC Berkeley
USGS topographic quadrangles	1893-present	USGS Bay Area Regional Database
USDA soil maps	1914-present	John Muir National Historic Site; U.S. Department of Agriculture
Wieslander VTM data	1933	Information Center for the Environment, UC Davis

Table 2. Summary of maps collected and reviewed.

County and regional surveys

Late 19th and early 20th century county and regional surveys were produced for a variety of purposes, and depict the study area at a relatively coarse scale. County and regional surveys collected for this study include 13 county maps from between 1871 and 1940, USGS topographic quads from [1893-4]1897 and [1893-4]1915, and USDA soil maps from 1914 and 1933.

The information provided in early county maps is limited to property boundaries, major infrastructure, and the general location of larger streams; vegetation cover is not shown (Fig. 5). Strentzel Creek, which drains the southern portion of Mt. Wanda, is depicted with varying degrees of detail on many of the county maps. Early USGS quadrangles likewise provide no information about vegetation cover, but show elevation contours, infrastructure, and larger streams. Strentzel Creek is shown as a dashed line disconnected from Alhambra Creek on the USGS maps, indicating that surface flow in Strentzel Creek was intermittent at the time of the surveys (Fig. 6).

Figure 4. Maps, photographs, and textual documents comprised the principal data types collected for this study. Left to right: a 1915 topographic map of the Mt. Wanda area; an 1885 photograph of Dr. John Strentzel with the Martinez Adobe in the background; an 1888 mortgage to Dr. Strentzel. (USGS [1893-4]1915; HABS CAL,7-MART,2--1, courtesy of Library of Congress, Prints & Photographs Division; County of Contra Costa 1888, John Muir Papers, Holt-Atherton Special Collections, University of the Pacific Library. ©1984 Muir-Hanna Trust)

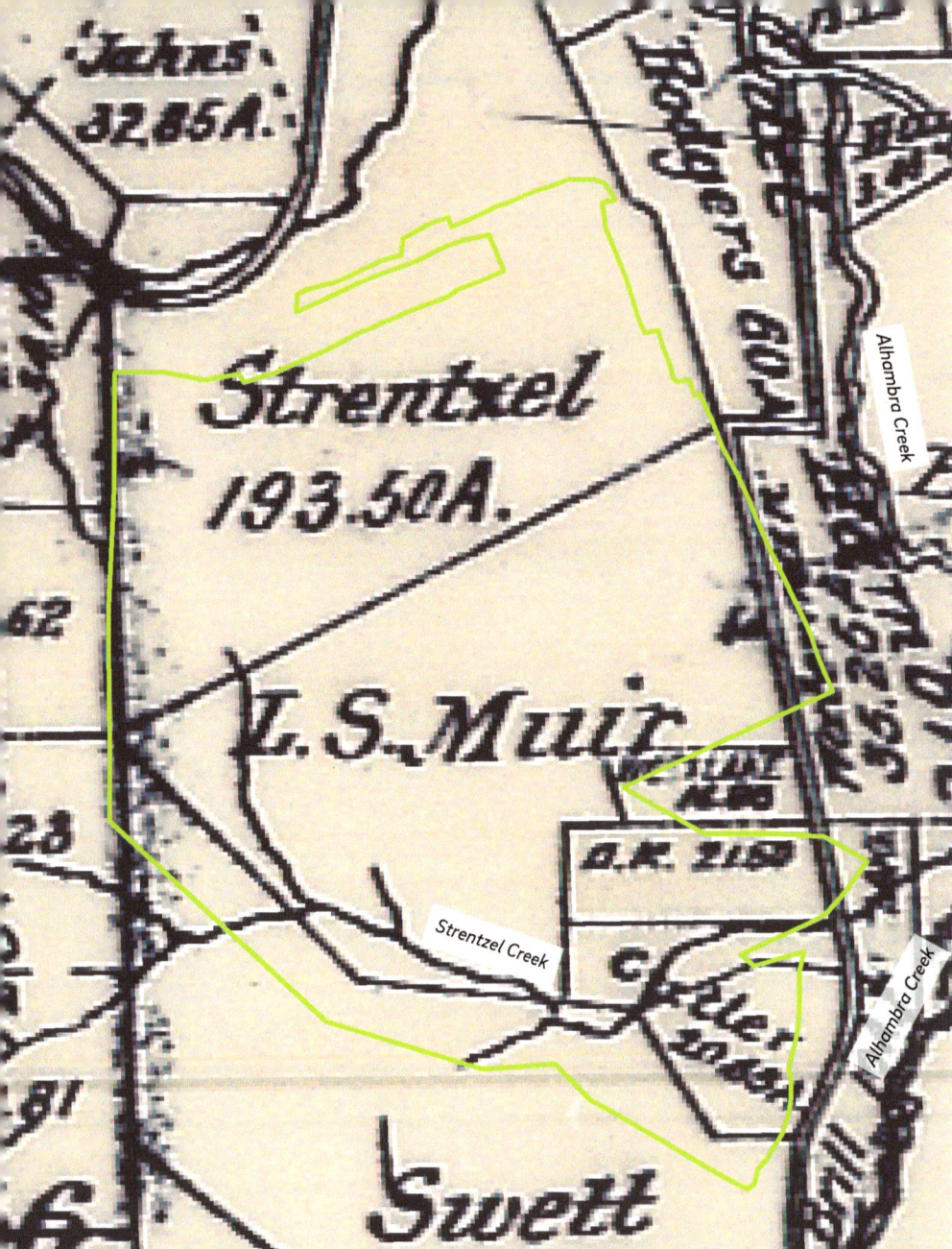

The earliest soil map covering Mt. Wanda is from 1914, though a more detailed mapping effort was conducted in 1933. Information available in historical soil maps and accompanying surveys, such as soil texture, depth, water-holding capacity, salinity, and presence or absence of a hardpan, often provides evidence for the historical occurrence of wetlands or particular vegetation communities. Soils are a relatively persistent feature, and thus historical soil maps – while not as early as other sources – can provide an important window into 19th century conditions. However, the small scale of early soil maps can be a major limitation: in the 1933 survey almost the entire study area is classified as one of two phases of a single soil type (Hugo clay), thus precluding the identification of any heterogeneity in edaphic or vegetation characteristics (Fig. 7). The description of Hugo clay in the accompanying soil survey is similarly generic, but does broadly support patterns depicted in other sources: "Scattered oak trees grow over a large part of the land, with grasses occupying the open spaces between the trees" (Carpenter and Cosby 1939).

N
1,000 feet
1:12,000 scale
SCALE FOR MAPS PAGES 10-11

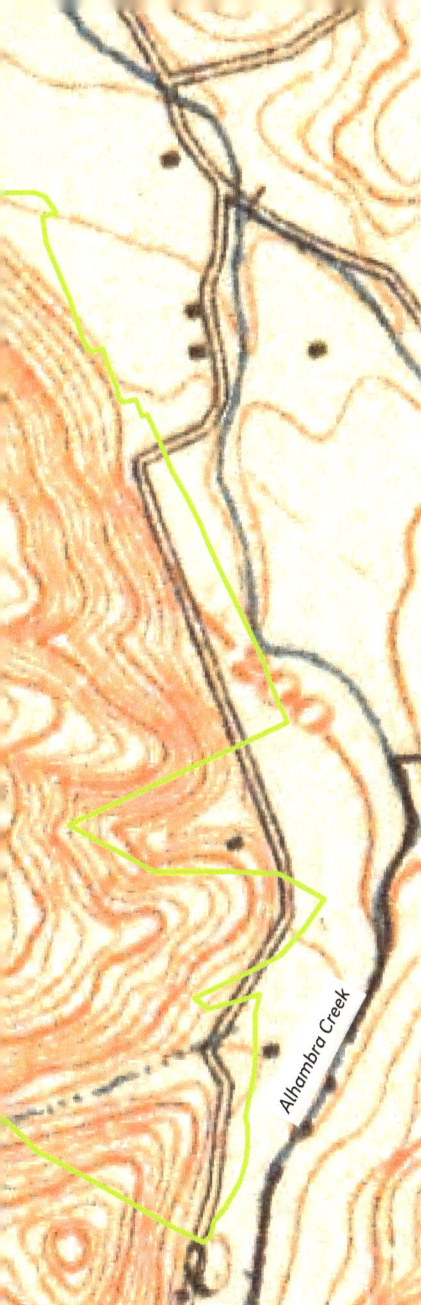

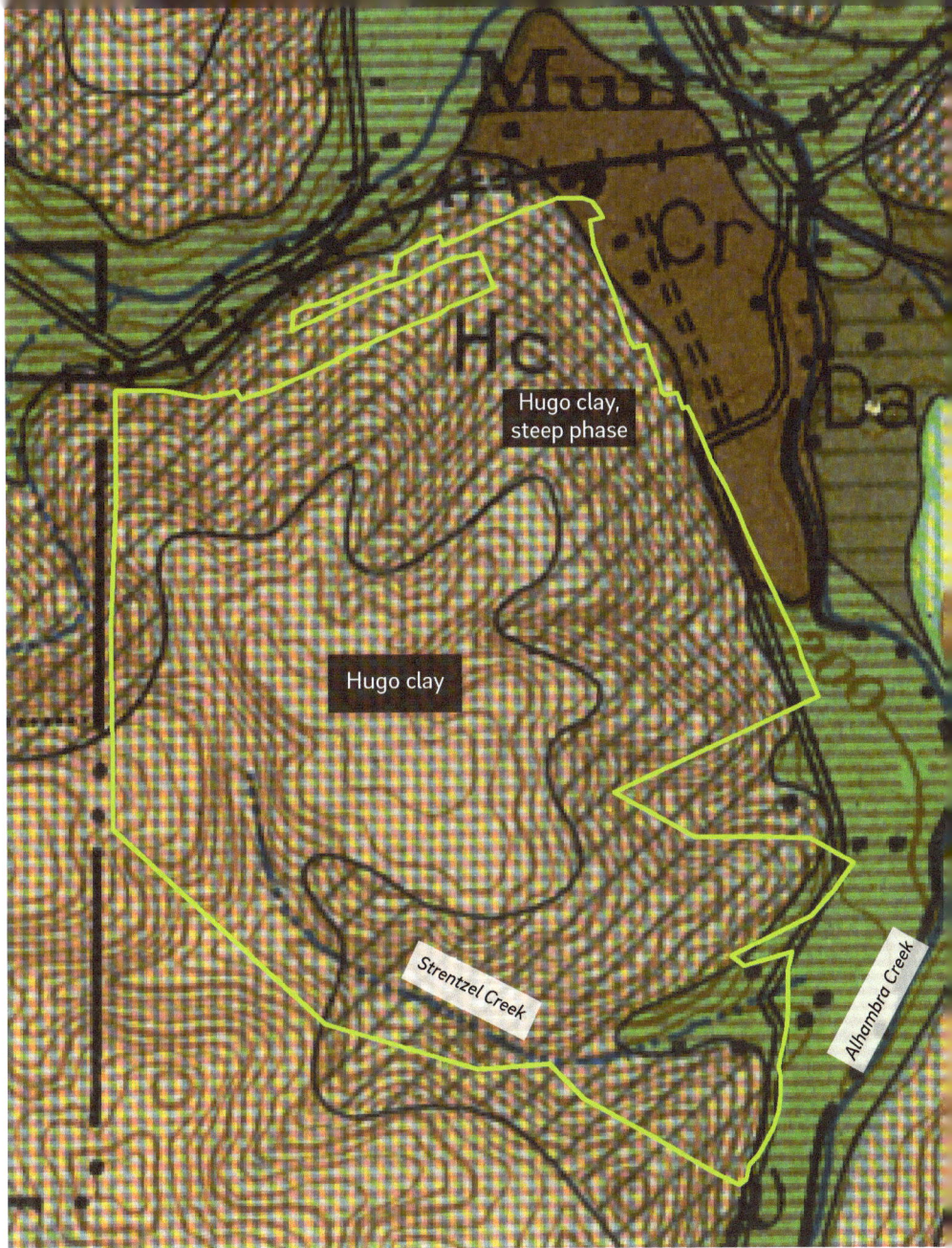

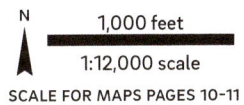

Figure 5. (left) Early county maps of the study area, such as this one from 1894, depict property boundaries, major roads, and larger streams such Strentzel Creek. Note that due to the coarse scale of the map the location of Strentzel Creek and other features appear somewhat offset relative to their true location. (Wagner and Sandow 1894, courtesy of Contra Costa County Public Works Department)

Figure 6. (center) Strentzel Creek is depicted as an intermittent stream on the southern end of the study area in the early USGS maps, and is shown as disconnected from Arroyo del Hambre (Alhambra Creek) to the east. (USGS [1893-4]1897)

Figure 7. (right) The 1933 soil map of Mt. Wanda classifies nearly all of the study area as one of two phases of Hugo clay. (Carpenter and Cosby 1933)

John. STRENTZEL ESTATE.

Mrs WOLFORD' Properties

Public Lands

VINEYARD
12½ Acres

J. & L. MUIR PROP'TY

PASTURAGE Land
PASTURAGE L.
100½ Acres

VINEYARD

82½ Acres

PASTURAGE Land

PUBLIC ROAD

John Swett's Land

VINEYARD 30 Acres

12½ Acres

PASTURAGE Land
27½ Acres

Orchard & Vine

'Orchard'
'VINEYARD'
15 Acres

Windmill

PINOLE RANCH

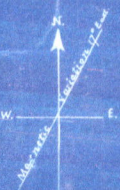

N
W — E
S

ALHAMBRA VALLEY

Strain or To

Damuir

PLAT
of a portion of the
(ALHAMBRA)
RANCH
Contra Costa Co
Cal
March 1891
E.C. Brown
Scale 3 Chains To One Inch

Finer-scale surveys

In addition to coarse-scale county and regional maps, local surveyors produced finer-scale maps of individual parcels and subdivisions. These maps often include more detailed depictions of creeks, springs, vegetation, or other aspects of the historical landscape, but their spatial coverage is inconsistent. We examined over 400 local survey maps from the Contra Costa County Clerk-Recorder and other source institutions, but almost all of the maps were located outside of the study area.

An 1891 survey map of the Alhambra Ranch (Brown 1891) provides some information about historical vegetation communities in the study area (Fig. 8). Much of the southern and western portions of Mt. Wanda are labeled as "pasturage." The term "pasturage" implies that grass cover was extensive enough to permit grazing in these areas, though the term is not synonymous with grassland and does not preclude the presence of substantial tree cover. Indeed, significant portions of the area labeled as pasturage are shown with forest or woodland cover in the 1939 aerial photos.

Figure 8. This 1891 map of Alhambra Ranch labels large portions of the southern and western side of study area as "pasturage," indicating that grass cover was sufficient to permit grazing. (Brown 1891, image courtesy of John Muir NHS, JOMU 3889)

Wieslander VTM mapping

The first spatially-comprehensive vegetation map of Mt. Wanda is the Wieslander VTM map produced in 1933. Albert Everett Wieslander was a forester who led a vegetation survey throughout large areas of the state in the 1920s and 1930s. Surveyors in the field observed vegetation cover from ridges and other high points and recorded their observations on USGS quadrangles (Kelly et al. 2005). Vegetation types were mapped as either pure or mixed stands of a single vegetation type (>80% herb, shrub, or tree cover) or as vegetation mosaics with a mixture of vegetation types. In pure or mixed stands, species with greater than 20% cover were recorded as dominants; in mosaics, species within each vegetation type with greater than 20% cover were recorded as dominants (Wieslander et al. n.d., Keeler-Wolf 2007).

The VTM map of Mt. Wanda shows grasslands occupying the central, higher-elevation portions of the study area, while oak woodlands are mapped along the periphery and in the northern part of the study area (Fig.

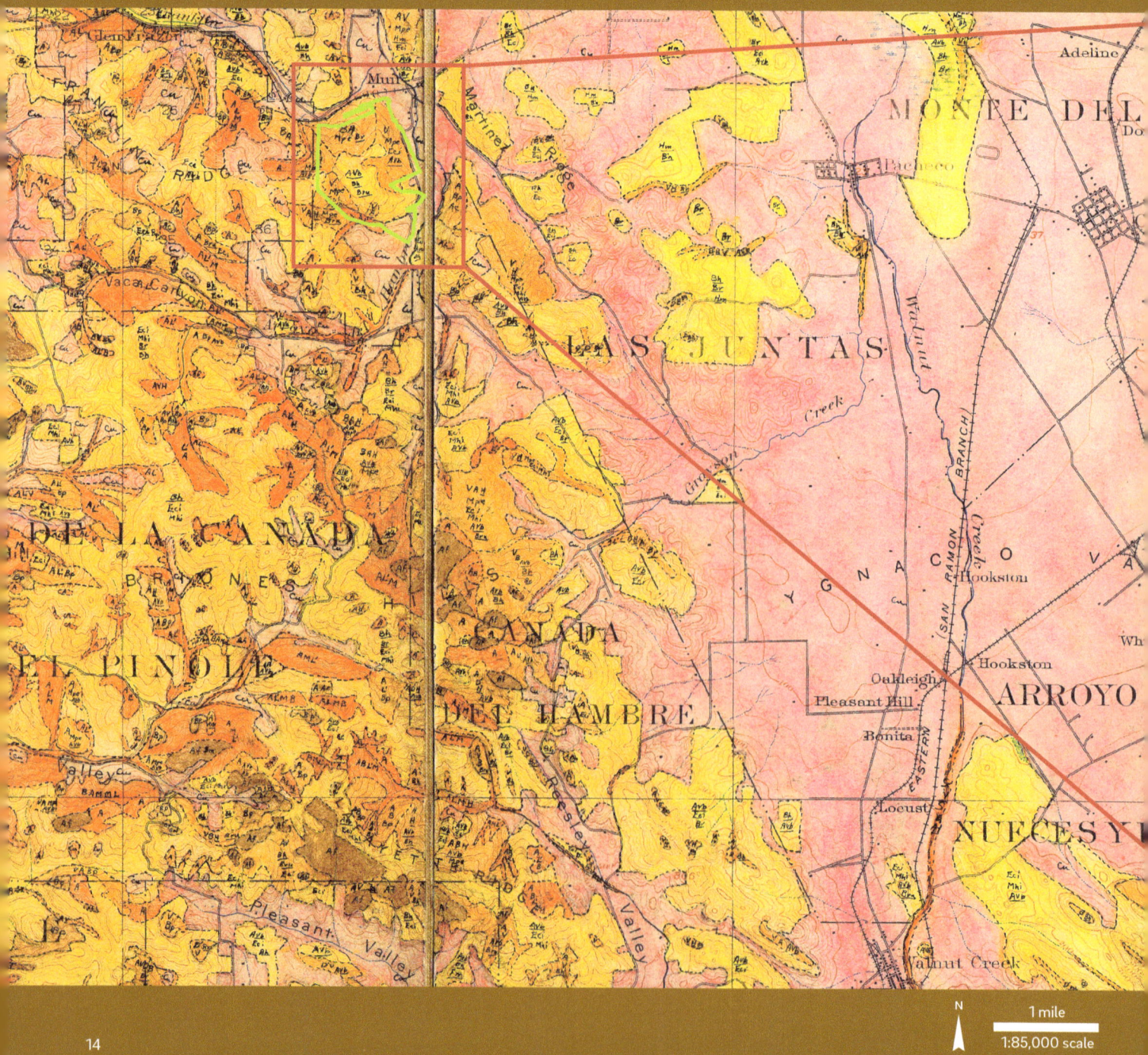

N

1 mile

1:85,000 scale

9). The dominant species recorded in the grasslands include three introduced species of European origin: slender oat *(Avena barbata),* soft brome *(Bromus hordeaceus),* and foxtail brome *(B. madritensis* ssp. *rubens)* (D'Antonio et al. 2007, DiTomaso et al. 2013).[1] In several large areas on the northeastern, western, and south- ern sides of Mt. Wanda, valley oak *(Quercus lobata)* is listed as a dominant overstory species in the VTM map, and in some of these areas coast live oak *(Q. agrifolia)* and buckeye *(Aesculus californica)* are also listed as dominants. Black oak *(Q. kelloggii)* and coast live oak are listed as dominant species in a large area on the northern side of Mt. Wanda, while coast live oak and California bay laurel *(Umbellularia californica)* are listed as dominants in a small area on the southern side of the study area. Blue oak *(Q. douglasii)* is not listed as a dominant tree species anywhere on the site in the VTM map.

---

[1] Slender oat was accidentally introduced to California as early as the late 18th century, while soft brome and foxtail brome arrived in the mid-19th century (Guma et al. 2006, D'Antonio et al. 2007, DiTomaso et al. 2013).

A	Quercus agrifolia
Avb	Avena barbata
B	Quercus kelloggii
Bh	Bromus hordeaceus
Br	Bromus diandrus
Bru	Bromus madritensis rubens
Cu	Cultivated
Eci	Erodium cicutarium
H	Aesculus californica
L'	Umbellularia californica
Mpe	Claytonia perfoliata
V	Quercus lobata

Figure 9. The 1933 Wieslander VTM map showing dominant plant species recorded on Mt. Wanda. (courtesy of Information Center for the Environment, UC Davis)

N
1,000 feet
1:15,000 scale

Photographs

Detailed information about the early Mt. Wanda landscape is provided in both landscape and aerial photographs (Table 3). We collected approximately 120 landscape photographs of Mt. Wanda, Alhambra Valley, and surrounding areas taken between ca. 1885 and present. We also consulted a photomosaic compiled from 1939 aerial photographs (the earliest available for the region; USDA 1939) as well as modern aerial photos (USDA 2009).

Source	Date Range	Source Institution(s)
Landscape photographs	ca. 1885-present	The Bancroft Library, UC Berkeley; Contra Costa County Historical Society; Holt-Atherton Special Collections, University of the Pacific Library; John Muir National Historic Site; Water Resources Collections and Archives
Aerial photographs	1939-2009	Earth Sciences & Map Library, UC Berkeley; U.S. Department of Agriculture, National Agriculture Imagery Program

Table 3. Summary of photographs collected and reviewed.

Landscape photos

Most of the early landscape photos we collected of the study area and were able to geolocate show either the northern side of Mt. Wanda near the Muir House or the eastern side along Alhambra Valley. We collected numerous additional photographs of Alhambra Valley and the surrounding hills, but these photographs either do not show the study area or lack sufficient information to determine their precise location.

Early landscape photos show forested areas along the northern and eastern slopes of Mt. Wanda adjacent to Alhambra Valley (Figs. 10-14). These appear to be primarily areas mapped as Coast Live Oak Forest and California Laurel Forest in the modern vegetation mapping (NPS 2004; see Fig. 24 on page 37). Identification of individual tree species from the historical photos cannot be

accomplished with full confidence, but multiple photos appear to show coast live oak, California bay, and California buckeye.

We were unable to locate any early photographs of grasslands that are unequivocally within the study area, though one ca. 1905 photo depicts a grassland on a hillside adjacent to Alhambra Valley that maybe include a portion of the study area (Fig. 15). The photograph appears to show a grassland dominated by annual grasses with several patches of shrubs; the presence of shrubs may indicate that this area was ungrazed or lightly grazed (Hunter et al. 1993).

Several photographs from the early 20th century appear to show a small patch of chaparral on the northern slope of Mt. Wanda near Franklin Canyon (Fig. 16). The chaparral patch extends beyond the frame of the photographs, however, and thus its size cannot be estimated. In a vegetation study from 1993, Hunter et al. (1993) state that chaparral was more extensive historically on this part of Mt. Wanda; they also claim that this patch of chaparral was burned between 1897 and 1902. Their main support for this statement is a single turn-of-the-century photograph (Li-55, Holt-Atherton Dept. Spec. Coll., University of the Pacific, Stockton), but we were unable to locate this photograph or to find other clear evidence of historical fires or changes in the extent of chaparral cover on Mt. Wanda.

Figure 10. (left) This ca. 1899 photograph is looking northwest across Alhambra Valley. The forested northeast slope of Mt. Wanda is visible to the left of the Alhambra Valley trestle. (photo #1082, courtesy of Contra Costa County Historical Society)

Figure 11. (center) This undated photograph (ca. 1905-25), taken from the eastern slope of Mt. Wanda, is looking north towards the Alhambra Valley trestle. The vegetation in the foreground appears to be relatively open mixed evergreen forest. The tree on the left side of the image appears to be an oak, and the tree just visible at the bottom of the image may be California bay. (BANC PIC 2011.049 2/2, #1049, courtesy of The Bancroft Library, UC Berkeley)

Figure 12. (right) A similar photo (undated), apparently taken a little lower on the slope, shows deciduous trees (possibly buckeye) in the foreground. (photo #2473, courtesy of Contra Costa County Historical Society)

Figure 13. (top left) This ca. 1890 photo shows the forested northern slope of Mt. Wanda behind the Muir House. A 1921 newspaper article about the property referred to "the wooded hills that climb up almost from the back porch of the home" (Cushing 1921). (Image courtesy of John Muir NHS, JOMU 1729)

Figure 14. (bottom left) Oaks are visible at the bottom of the northern slope of Mt. Wanda in this 1910 photograph, taken from the orchards near the Muir House. (Image courtesy of John Muir NHS, JOMU 4174)

Figure 15. (bottom center) This ca. 1905 photograph shows a grassland and patches of shrubs on a hillside (possibly Mt. Wanda) adjacent to Alhambra Valley. (photo #F14-0755, John Muir Papers, Holt-Atherton Special Collections, University of the Pacific Library. ©1984 Muir-Hanna Trust)

Figure 16. (right, top and bottom) These two photographs show comparable views of the Muir ranch, Alhambra Valley trestle, and northern slope of Mt. Wanda. The top photograph, dated ca. 1905 (Killion 2005), appears to show dense chaparral vegetation cover in the upper right corner of the photo. The bottom photograph, dated 1915, appears to show the same area with much sparser vegetation cover, though image quality may explain part of the difference. The reported dates for these photos are questionable: though ostensibly taken a decade later, the bottom photo appears to show the same orchards as the top photo in an earlier stage of growth, suggesting that it was taken first. (photo #F13-0645, John Muir Papers, Holt-Atherton Special Collections, University of the Pacific Library. ©1984 Muir-Hanna Trust; photo #2477, courtesy of Contra Costa County Historical Society)

Aerial photos

Though substantially later than other sources acquired in this reconnaissance study, the 1939 aerial photographs provide an extremely accurate picture of landscape-scale vegetation patterns that cannot be discerned from earlier sources (Fig. 17). The higher-elevation portions of the study area, including the Wanda and Helen summits, can be seen covered in grasslands with very little tree cover. Much of the grassland vegetation is located in an area with Los Osos clay loam soil type (Soil Survey Staff 2014). Forests and woodlands are visible in much of the northern half of the study area and in Strentzel Canyon on the southern side of the mountain. Forest density appears to be highest on the north-facing slopes.

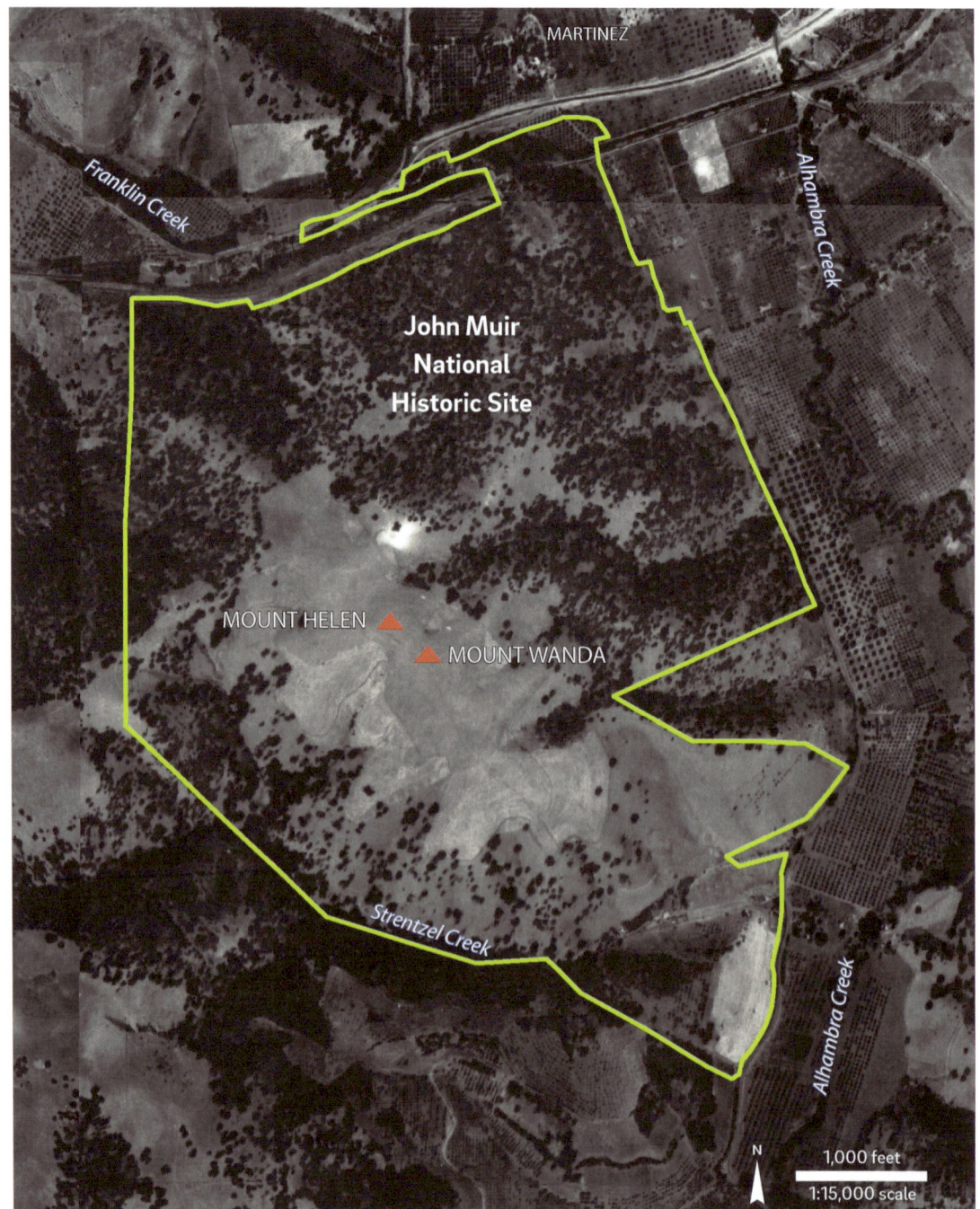

Figure 17. 1939 aerial photos of the study area. (USDA 1939)

Texts

Textual data was derived from a variety of sources, shown in Table 4. The earliest sources include observations from 18th century Spanish explorers and records from mid-19th century land grant court cases and land surveys. Mid- to late 19th and early 20th century sources include journal entries, letters, newspaper articles, species records, county histories, and books. The textual sources that we drew on most heavily were a series of 1895-6 journal entries by John Muir.

Some of the earliest written descriptions of the general area come from journal entries by Spanish explorers traveling along the southern shore of the Carquinez Strait, including Pedro Fages and Juan Crespí in March 1772 and Juan Bautista de Anza in August 1776. Though neither party passed by Mt. Wanda, both camped near present-day Martinez; their descriptions of the landscape suggest a mosaic of grassland, oak savanna, and woodland/forest. In March of 1772, for instance, Fages wrote, "All the land we crossed up to here was very level and grassy and excellent texture; there were a good many white-oaks and live-oaks studded all over it" (Fages and Treutlein 1972). In April of 1776, de Anza noted that the "canyons are well provided" with "timber of oak and live oak" (de Anza and Bolton 1930).

The richest data sources located for this study are late 19th and early 20th century journal entries by John Muir, his daughter Helen Muir, his mother-in-law Louisiana Strentzel, and other friends and relatives, as well as letters to and from John Muir. Muir's journal entries from 1895-6 represent an especially valuable source of information, providing a detailed record of the plants, birds, and other species that he observed on his walks up Mt. Wanda and nearby hills. Though these observations lack the level of spatial precision often associated with maps and photographs, they provide important insights into historical species presence and abundance, vegetation structure, land and water uses, and the timing of natural history events such

Source	Date Range	Source Institutions
Spanish explorer diaries	1772-6	California Historical Society; UC Berkeley
Land surveys, deeds, and land grant court case documents	1837-91	The Bancroft Library, UC Berkeley; Holt-Atherton Special Collections, University of the Pacific Library
Journal entries, autobiographies	1850-1902	The Bancroft Library, UC Berkeley; Holt-Atherton Special Collections, University of the Pacific Library; John Muir National Historic Site
General Land Office field notes	1852-83	Bureau of Land Management
Letters	1875-1908	Holt-Atherton Special Collections, University of the Pacific Library
General or county histories	1879-82	Contra Costa County Library
Species records	1887-1933	Consortium of California Herbaria
Newspaper articles	1889-1922	California Digital Newspaper Collection; Holt-Atherton Special Collections, University of the Pacific Library
Books, theses, technical reports	1912-2014	The Bancroft Library, UC Berkeley; Contra Costa County Library; John Muir National Historic Site; Online databases; Shields Library, UC Davis; Water Resources Collections and Archives

Table 4. Summary of textual data collected and reviewed.

as flowering or nesting. Several other types of textual sources provided useful data about the early Mt. Wanda landscape, including land grant court case documents, General Land Office field notes, newspaper articles, and species records.

Early texts document three species of deciduous oak trees – blue oak *(Quercus douglasii),* valley oak *(Q. lobata),* and black oak *(Q. kelloggii)* – growing on Mt. Wanda or the surrounding hills. John Muir made numerous references to all three species in journal entries from 1895-6 (blue oak is also referred to as "Douglas" oak; valley oak as "white oak" or "lobata;" and black oak as "Kellogg oak"):

> 1895, March 30: "The Kellogg Oak with yellow-green foliage is now nearly in prime. The first to put forth its leaves. The Douglas + lobata also yellow-green + purplish in foliage but less lively in color + later in putting forth leaves – they are not yet full grown" (Muir 1895)[1]

> 1896, Feb 24: "A week to ten days difference in time of flowering of Kellogg Oak. Some lvs [leaves] two inches long." (Muir 1896)

> 1896, Feb 27: "Some of the live + white oaks opening buds. One white oak has leaves nearly unfolded must have buds close" (Muir 1896)

> 1896, March 8: "In the evening Helen + I took a walk in the West hills. Some of the live oaks just opening buds. Other with shoots six inches long. So also the blue oaks. + White-oaks. The different tones of brown + purple + yellow green swelling in rounded broad masses on the hills very fine." (Muir 1896)

> 1896, March 12: "The oaks with varying tones of yellow brown + green making delightful pictures on the hills. A few of the blue oaks not yet in leaf – buds just beginning to open, a few are open + male flrs [flowers] in full fringe." (Muir 1896)

Though the historical distribution and extent of each species cannot be determined from the available data, a letter from the botanist George Engelmann (for whom the Engelmann oak is named) to John Muir suggests that blue oak was more abundant than valley oak on the hills adjacent to Muir's Alhambra ranch house:

> "As I certainly shall not (if at all) be there in season for the flowering of the oaks, I would ask you to get for me at least the two species which grow close about you. One is Quercus lobata, which grows as you know, above the stables on the Creek; the other is Quercus Douglasii, the blue mountain oak, on the bare hills above you. A few specimens of each of them in flower and again a week or two later when the leaves are not yet fully grown (showing the female flowers or young acorns) would be very desirable. I would say that I found a few specimens of the lobata also on the hills, but Douglasii is readily distinguished by the smoother, whiter bark, the smaller and less lobar leaves; it seems to be the common tree on the arid hills." (Engelmann 1881)

Coast live oak (*Q. agrifolia*), California bay (also known as California laurel, *Umbellularia californica*), and California buckeye (*Aesculus californica*) – dominant trees in coast live oak forests and other mixed evergreen forests in this region – were documented in a variety of historical textual accounts. In a journal entry from 19 March, 1872, for instance, Louisiana Strentzel writes, "The sun came shining down aslant the hills, throwing a flood of light over the pale green slopes, and the dark rich green of the buckeyes, laurels and live oaks" (Strentzel 1850-82). In his biography, John Strentzel recalls arriving to Alhambra Valley and observing "the hills and valley partially covered with magnificent laurel, live- oak and white- oak trees" (Strentzel

---

[1] It is possible that entries which combine two species names (e.g., "Douglas + lobata") reflect Muir's recognition of hybridization between oak species.

n.d.). In March, 1895, John Muir noted, "The pea green foliage of buckeye has fine effect" (Muir 1895). Buckeyes would likely have occurred on Mt. Wanda in both deciduous oak woodlands and mixed evergreen forest; there are also records of Muir planting buckeyes near the vineyards (Killion 2005).

Several late 19th century sources appear to describe mixed evergreen forests and oak woodlands in and around Strentzel Canyon on the southern side of the study area. In 1892-3, John Muir and Louie Strentzel Muir sold a portion of their land on the southeastern side of Mt. Wanda to Roger Cutler, and the descriptions of the parcels in the accompanying deeds document several species of trees near lower Strentzel Creek, including live oaks, valley oaks ("white oak[s]"), and a buckeye; another valley oak is documented just over a small ridge further to the southeast (County of Contra Costa 1892, 1893). In addition, a newspaper article from 1889 describes the hills around John Swett's property, just to the south of the study area, as "covered with evergreen trees, except in portions where the oaks have been leveled and their places taken by the choicest vines, deciduous fruits and olive trees" (Pacific Rural Press 1889).

California black walnut *(Juglans californica* var. *hindsii),* which occurs today in riparian areas on the northern side of Mt. Wanda (Jepsen and Murdock 2002), may have occurred historically within the study area: in a 1908 letter to John Muir, the botanist Charles Sprague Sargent requests a specimen of "California Walnut which, as you Know, does not grow very far from your house" (Sargent 1908). Several other riparian species, including willows (*Salix* spp.), which are found today on the north slope of Mt. Wanda, and Fremont cottonwood (*Populus fremontii* ssp. *fremontii*), which occurs on the southeastern side of the parcel (Jepson and Murdock 2002), are also mentioned in historical descriptions of the Mt. Wanda area (Strentzel 1850-82, Muir 1895, Muir 1896).

Grasslands, in some areas supporting sparse oaks, were documented in a variety of textual sources. While traveling north through the study area in 1852, for instance, General Land Office surveyor Robert B. Hays (1852) noted "very little scattering oak timber" on the western side of the area mapped as grassland today. This field note was recorded at a section corner, however, and thus refers to the general character of the landscape extending one mile to the south, rather than to local conditions at this specific point.

By the time that John Muir settled in the Alhambra Valley, the grasslands of Mt. Wanda, like grasslands throughout the state, had already been transformed by the spread of introduced annual grasses. Wild oats *(Avena* spp.) were among the most common species: John Strentzel wrote that upon arriving in Alhambra Valley there was "everywhere a green mantle of wild oats from 8-12 inches high" (Strentzel n.d.). In April 1868, Louisiana Strentzel described her "little children trying to wade through the wild oats which were at that time over one foot high" (Strentzel 1850-82), and in February 1881 she wrote, "Hills covered with grass several inches high" (Strentzel in Killion 2005). John Muir admired the oats in a journal entry from 8 May, 1896:

> "Walked over the hills with Wanda + Helen. How the wind did surge... + how the wild oats danced + rippled + clapped their spikelets like happy hands in a passion of joy." (Muir 1896)

Unlike the native grasses, many of which were perennial, the introduced annual grasses died back during the summer time. In July of 1895, for instance, Muir wrote, "The pastures are all overrun with Chili Thistle. Scarce any grasses left. Now dry + dead" (Muir 1895). The conversion from native perennial grasses to introduced annual grasses likely had many implications for wildlife and hydrology. Moore (2006), for instance, proposes that the combination of decreased root depth and mass and increased wet season soil moisture resulting from the invasion of annual grasses could be a primary driver of ongoing channel incision on Mt. Wanda.

John and Helen Muir's accounts of the Mt. Wanda landscape and surrounding areas from 1895-1902 include observations of many species of flowering forbs and shrubs (Fig. 18). These records provide a colorful window into the botanical diversity of the historical landscape, and in some cases even provide information about historical species abundance or population trends. In April of 1895, for instance, John Muir wrote, "Found one nemophila 'baby-blue-eye'. Nearly extinct hereabouts – once abundant – so also most of the gilias. Found one large patch of orthocarpus purple – also white + yellow." (Muir 1895). In most cases the journal entries do not provide precise information about the locations of the observed plants, but in several cases the entries are more spatially explicit. On April 2, 1896, for example, John Muir recorded "A 2nd species of Wythea near summit of Helen hill" (Muir 1896), and on February 27, 1902, Helen Muir observed "a little bunch of butter-cups" and "some lovely Nuttalia" near the railroad tunnel on the northern side of Mt. Wanda (Muir 1901-2).

While the historical textual accounts compiled for this study provide a wealth of detailed information about the historical vegetation community composition of Mt. Wanda and the surrounding areas, identifying which species are described in these accounts is often challenging. In many cases, the taxonomic terminology used in historical references is ambiguous or antiquated, and thus determining a single modern species equivalent is not possible. In these cases, the best that can be done is to identify several likely candidate species based on modern species distribution. To facilitate the interpretation of botanical references in historical accounts, we used modern species inventories for the John Muir National Historic Site to identify possible modern equivalent species for plants described in early records (Table 5).

In addition to recording the names of observed plants, John Muir and others also took notes about various natural history events that they observed on Mt. Wanda and in surrounding areas, such as when oaks and other trees were leafing out, when wildflowers were in bloom, and when birds were building nests (Tables 6 and 7). These notes provide valuable insights into the life cycles of numerous species and the seasonal changes that were taking place in the landscape nearly 120 years ago. The timing of natural history events – which is the focus of the field of study known as phenology – is an area of much current scientific interest, in large part because climate change has begun to cause shifts in the timing of critical life cycle events for many species. The John Muir National Historic Site, as a collaborator on the California Phenology Project (www.usanpn.org/cpp/node/8), is currently monitoring phenological trends for target species; comparing this contemporary information with the historical observations could contribute to an understanding of long-term species responses to climate change.

The historical phenology data compiled for this study have a number of limitations that make comparison with modern phenology data challenging. First, the historical observations are sometimes ambiguous with regard to taxonomy. Second, while all of the observations were made in the general area around Mt. Wanda and the ranch, the precise location is not known in most cases; in fact, most of the accounts represent generalized, population-level descriptions rather than observations about individual plants. Third, the descriptions of the life history events, or "phenophases," are often informal or vague, and do not adhere to standardized definitions of phenophases used by modern researchers (e.g., Denny et al. 2014). Fourth, historical accounts typically focus on the presence or initiation of the most notable phenophases and lack information about the presence, absence, or intensity of other phenophases. Finally, the historical dataset only spans two years, and thus the events that Muir observed may have been heavily influenced by climatic anomalies in those years. Nevertheless, the dataset contains early phenological observations for a wide range of species in the Mt. Wanda area, and could be of considerable use as baseline information for comparison with modern phenology data. Only historical observations related to plant phenology are presented here, though the textual data also includes references to the timing of life cycle events for several animal species (see Appendix).

Additional information about the historical vegetation composition on Mt. Wanda can be found in Cummings and Puseman (1991), who conducted a pollen, phytolith, and macrofloral analysis of a single adobe brick from the Martinez Adobe. The results of their analysis reveal numerous plant species or groups that were present in the surrounding area at the time that the Martinez Adobe was constructed in 1849, complementing the historical data collected as part of this reconnaissance study. The most abundant plant remains found in the brick were from Festucoid grasses (e.g., *Avena, Festuca*) and the High-spine Compositae (Asteraceae) group (e.g., aster, rabbitbrush, snakeweed, sunflower). Less abundant remains were found for many

---

Table 5. (following pages) Plants described on or near Mt. Wanda in historical documents, and possible modern equivalents. The terms in the Historical Observation column have been transcribed verbatim from the original sources (in some cases minor variations have been omitted), and do not necessarily conform to any accepted modern names. Probable references to ornamental or cultivated plants restricted to gardens or orchards around the Muir/Strentzel properties were not included. To compile the list of Possible Modern Equivalent(s), we searched for the historical observation (including spelling variations) in the 2002 JOMU vegetation inventory (Jepsen and Murdock 2002) and the JOMU Species List (NPS 2014). We added a species to the list of possible modern equivalents if the historical observation matched the common name or Latin name of a species listed in Jepsen and Murdock (2002) as occurring on or potentially found on Mt. Wanda. Species currently listed as "rejected" for Mt. Wanda were also added to the list of possible modern equivalents if they plausibly could have occurred in the area in the past. In several instances where the historical observations appeared to correspond to a family name, we added possible matches within that family. Additional species that were listed in the JOMU Species List (not limited to Mt. Wanda) but not listed in Jepsen and Murdock (2002) were also added. In cases where the historical observation had no matching equivalent in Jepsen and Murdock 2002 or NPS 2014 (e.g., "Orthocarpus"), we used Calflora (http://www.calflora.org/) to identify alternate or superseding names.

* = non-native

[1] = only listed in Jepsen and Murdock 2002

[2] = only listed in NPS 2014

[3] = not listed in Jepsen and Murdock 2002 or NPS 2014

No footnote = listed in both Jepsen and Murdock 2002 and NPS 2014

Historical Observation	Source	Possible Modern Equivalent(s)
"Agrifolia"	Muir 1896 (27-Feb)	Coast live oak (*Quercus agrifolia*)
"Aliums"	Muir 1895 (12-Apr)	Serrated onion (*Allium serra*) White garlic (*Allium neapolitanum*)*[2]
"Blue oak" / "Quercus douglasii"	Muir 1895 (30-Mar), Muir 1896 (18-Feb, 8-Mar, 12-Mar)	Blue oak (*Quercus douglasii*)
"Brodiaeas"	Muir 1895 (12-Apr)	Harvest brodiaea (*Brodiaea elegans* ssp. *elegans*) Brodiaea (*Brodiaea coronaria*)[1] Elegant brodiaea (*Brodiaea elegans*)[2] Dwarf brodiaea (*Brodiaea terrestris*)[2]
"Buckeye"	Strentzel 1850-82 (19-Mar-1872), Muir 1895 (26-Jan, 30-Mar), Muir 1896 (16-Feb)	California buckeye (*Aesculus californica*)
"Bush mimulus"	Muir 1895 (18-Jul)	Sticky monkey flower (*Mimulus aurantiacus*)
"Buttercups"	Muir 1895 (24-Jan, 12-Apr), Muir 1896 (14-Feb, 31-Mar), Muir 1902 (27-Feb)	California buttercup (*Ranunculus californicus*) Spiny buttercup (*Ranunculus muricatus*)* Water buttercup (*Ranunculus aquatilis*) Delicate buttercup (*Ranunculus hebecarpus*)
"California Walnut"	Sargent 1908	California black walnut (*Juglans californica* var. *hindsii*)
"Castilleia"	Muir 1896 (16-Feb)	Indian paintbrush (*Castilleja affinis* ssp. *affinis*) Valley tassels (*Castilleja attenuata*) Purple owl's clover (*Castilleja exserta* ssp. *exserta*) Woolly owl's clover (*Castilleja foliolosa*) Cream sacs (*Castilleja rubincundula* ssp. *lithospermoides*)
"Chickweed"	Muir 1895 (24-Jan)	Chickweed (*Stellaria media*)* Sticky chickweed (*Cerastium glomeratum*)*
"Chili thistle"	Muir 1895 (18-Jul)	Unknown
"Clovers"	Muir 1895 (12-Apr, 2-May)	Many
"Collinsia sparsiflora"	Muir 1896 (27-Mar)	Blue-eyed mary (*Collinsia sparsiflora* var. *sparsiflora*)
"Cottonwood"	Muir 1895 (18-Feb)	Fremont cottonwood (*Populus fremontii* ssp. *fremontii*)
"Dodecatheon"	Muir 1895 (24-Jan, 12-Apr), Muir 1896 (18-Feb, 24-Feb)	Shooting star (*Dodecatheon hendersonii*)
"Gilias"	Muir 1895 (12-Apr)	Purplespot gilia (*Gilia clivorum*)[2]
"Gnaphalium"	Muir 1895 (18-Jul)	California everlasting (*Gnaphalium californicum*) Cudweed (*Gnaphalium luteo-album*)*
"Hosackia"	Muir 1895 (18-Jul)	Unknown
"Kellogg Oak"	Muir 1895 (30-Mar), Muir 1896 (18-Feb, 24-Feb)	Black oak (*Quercus kelloggii*)
"Larkspurs"	Muir 1895 (Apr-12)	Foothill larkspur (*Delphinium hesperium* ssp. *hesperium*)[1] Zigzag larkspur (*Delphinium patens*)[2]
"Laurel"	Strentzel n.d. (1853), Strentzel 1850-82 (19-Mar-1872), Muir 1895 (24-Jan), Muir 1896 (14-Feb, 1-Mar)	California bay (*Umbellularia californica*)
"Live oak"	Strentzel n.d. (1853), Strentzel 1850-82 (19-Mar-1872), Muir 1895 (30-Mar), Muir 1896 (27-Feb, 1-Mar, 8-Mar)	Coast live oak (*Quercus agrifolia*) Interior live oak (*Quercus wislizeni*)[1]

Historical Observation	Source	Possible Modern Equivalent(s)
"Monardella"	Muir 1895 (18-Jul)	Coyote mint (*Monardella villosa* ssp. *villosa*)
"Mustard"	Muir 1895 (24-Jan)	Black mustard (*Brassica nigra*)* Field mustard (*Brassica rapa*)* Hoary mustard (*Hirschfeldia incana*)* Hedge mustard (*Sisymbrium officinale*)* Tower mustard (*Arabis glabra* var. *glabra*)
"Nemophila"	Muir 1895 (24-Jan)	Fivespot (*Nemophila maculata*) Baby blue-eyes (*Nemophila menziesii* var. *menziesii*)[1] Canyon nemophila (*Nemophila heterophylla*)[2]
"Nemophila 'baby-blue-eye'"	Muir 1895 (28-Apr)	Baby blue-eyes (*Nemophila menziesii* var. *menziesii*)[1]
"Nuttalia"	Muir 1896 (14-Feb), Muir 1901-2 (27-Feb)	Oso berry (*Oemleria cerasiformis*)
"Orthocarpus purple – also white + yellow"	Muir 1895 (28-Apr)	Dwarf owl's-clover (*Triphysaria pusilla*) Valley tassels (*Castilleja attenuata*) Purple owl's clover (*Castilleja exserta ssp. exserta*) Cream sacs (*Castilleja rubincundula* ssp. *lithospermoides*) Butter-and-eggs (*Triphysaria eriantha* ssp. *eriantha*)[1]
"Portulaccas"	Muir 1895 (25-Jan, 12-Apr)	Red maids (*Calandrinia ciliata*) Miner's lettuce (*Claytonia perfoliata* ssp. *perfoliata*)
"Quercus lobata"/"Lobata"	Engelmann 1881, Muir 1895 (30-Mar)	Valley oak (*Quercus lobata*)
"Ribes"	Muir 1896 (14-Feb)	Hillside gooseberry (*Ribes californicum* var. *californicum*) Fuchsia flowered gooseberry (*Ribes speciosum*)[2]
"Saxifrage"	Muir 1896 (24-Feb)	California saxifrage (*Saxifraga californica*) Common woodland star (*Lithophragma affine*) Hillside woodland star (*Lithophragma heterophyllum*) Bolander's woodland star (*Lithophragma bolanderi*)
"Soap root"	Muir 1896 (16-Feb)	Wavy-leaved soap plant (*Chlorogalum pomeridianum* var. *pomeridianum*)
"Sycamore"	Strentzel 1850-82 (19-Mar-1872), Muir 1896 (29-Feb)	California sycamore (*Platanus racemosa*)[2]
"Valerianella"	Muir 1896 (31-Mar)	Seablush (*Plectritis congesta*) Long-spurred seablush (*Plectritis ciliosa* ssp. *ciliosa*) Short-spurred plectritis (*Plectritis brachystemon*)[2]
"White oak"	Strentzel n.d. (1853), Muir 1896 (27-Feb, 1-Mar, 8-Mar, 27-Mar)	Valley oak (*Quercus lobata*) Oregon white oak (*Quercus garryana*)[2]
"Wild oats"	Strentzel n.d. (1853), Strentzel 1850-82 (5-Apr-1868), Muir 1896 (8-May)	Slender wild oat (*Avena barbata*)* Wild oat (*Avena fatua*)*
"Willows"	Strentzel 1850-82 (19-Mar-1872), Muir 1895 (24-Jan)	Red willow (*Salix laevigata*) Arroyo willow (*Salix lasiolepis*)
"Wythea"	Muir 1896 (12-Apr)	Narrowleaf mule ears (*Wyethia angustifolia*) Grey mule ears (*Wyethia helenoides*)

Figure 18. Images of wildflowers representing species documented (or possibly documented) in Muir's journals and other sources. (clockwise from upper left: *Claytonia perfoliata*, photo by Miguel Vieira; *Gnaphalium californicum,* photo by M. Dolly; *Mimulus aurantiacus,* photo by Joyce Cory; *Dodecatheon hendersonii,* photo by Tom Hilton; *Nemophila menziesii,* photo by David O.)

Historical Observation	Date	Excerpt	Phenophase
"Aliums"	Apr 12, 1895	"in flower"[1]	Flowers or flower buds
"Brodiaeas"	Apr 12, 1895	"in flower"[1]	Flowers or flower buds
"Bush mimulus"	Jul 18, 1895	"still in flower"[1]	Flowers or flower buds
"Buttercup" / "Buttercups"	Jan 24, 1895	"in flower"[1]	Flowers or flower buds
	Apr 12, 1895	"in flower"[1]	Flowers or flower buds
	Feb 14, 1896	"in flower"[2]	Flowers or flower buds
"Castilleja"	Feb 16, 1896	"in flower"[2]	Flowers or flower buds
"Chickweed"	Jan 24, 1895	"in flower"[1]	Flowers or flower buds
"Clovers"	Apr 12, 1895	"in flower"[1]	Flowers or flower buds
"Collinsia sparsiflora"	Mar 27, 1896	"in flr"[2]	Flowers or flower buds
"Dodecatheon"	Jan 24, 1895	"in bud"[1]	Flowers or flower buds
	Feb 18, 1896	"Dodecatheon [in?] flr"[2]	Flowers or flower buds
	Feb 24, 1896	"in flr"[2]	Flowers or flower buds
"Gilias"	Apr 12, 1895	"in flower"[1]	Flowers or flower buds
"Gnaphalium"	July 18, 1895	"still in flower"[1]	Flowers or flower buds
"Grass" / "Grasses"	Jul 18, 1895	"Scarce any grasses left. Now dry + dead."[1]	N/A
	Mar 17, 1896	"Grass growing fast"[2]	Initial growth?
"Hosackia"	Jul 18, 1895	"still in flower"[1]	Flowers or flower buds
"Larkspurs"	Apr 12, 1895	"in flower"[1]	Flowers or flower buds
"Monardella"	Jul 18, 1895	"still in flower"[1]	Flowers or flower buds
"Mustard"	Jan 24, 1895	"in flower... gone to seed some of it"[1]	Flowers or flower buds; Fruits
"Nemophila"	Jan 24, 1895	"in flower"[1]	Flowers or flower buds
"Nemophila 'baby-blue-eye'"	Apr 28, 1895	"The flrs mostly gone to seed. Found one nemophila 'baby-blue-eye.'"[1]	Flowers or flower buds
"Nuttalia"	Feb 14, 1896	"in flower"[2]	Flowers or flower buds
"Orthocarpus purple – also white + yellow"	Apr 28, 1895	"Found one large patch of orthocarpus purple – also white + yellow"[1]	Flowers or flower buds
"Portulaccas"	Apr 12, 1895	"in flower"[1]	Flowers or flower buds
"Saxifrage"	Feb 24, 1896	"in flr"[2]	Flowers or flower buds
"Soap-roots"	Feb 16, 1896	"in flower"[2]	Flowers or flower buds
"Wild oats"	May 8, 1896	"spikelets"[2]	Flower heads

Table 6. Historical observations related to the timing of natural history events for forbs and grasses in and around the study area, and corresponding "phenophases" as described by Denny et al. 2014. The terms in the Historical Observation column have been transcribed verbatim from the original sources (in some cases minor variations have been omitted), and in many cases could refer to multiple different species (see Table 5). The events in the Phenophase column correspond to standardized phenophases for different plant functional groups as defined in Denny et al. 2014 (N/A indicates that a corresponding phenophase does not exist).

[1] = Muir 1895

[2] = Muir 1896

Historical Observation	Date	Excerpt	Phenophase
Black oak ("Kellogg oak," "Cal oak")	Mar 30, 1895	"yellow-green foliage is now nearly in prime"[1]	Leaves
	Feb 18, 1896	"full flower"[2]	Flowers or flower buds; Open flowers
	Feb 24, 1896	"some lvs two inches long"[2]	Leaves
Blue oak ("Douglas" [oak])	Mar 30, 1895	"yellow-green + purplish in foliage... leaves... not yet full grown"[1]	Increasing leaf size
	Feb 18, 1896	"buds not open yet"[2]	Breaking leaf buds (absence)
	Mar 8, 1896	"Some of the live oaks just opening buds. Other with shoots six inches long. So also the blue oaks. + White-oaks."[2]	Breaking leaf buds
	Mar 12, 1896	"A few... not yet in leaf – buds just beginning to open, a few are open + male flrs in full fringe"[2]	Breaking leaf buds; Leaves; Flowers or flower buds; Open flowers?
California bay	Jan 24, 1895	"full flower"[1]	Flowers or flower buds; Open flowers
	Feb 14, 1896	"in flower"[2]	Flowers or flower buds
California buckeye	Jan 26, 1895	"leaves opening rapidly... even the buds make a fine show"[1]	Breaking leaf buds
	Mar 30, 1895	"pea green foliage"[1]	Leaves
	Feb 16, 1896	"will soon be in full leaf"[2]	Increasing leaf size?
California sycamore	Mar 19, 1872	"young leaves"[3]	Increasing leaf size
	Feb 29, 1896	"just opening lvs + hanging forth their strings of brown heads of flrs. Leaves some of them [sic] nearly an inch long."[2]	Breaking leaf buds; Increasing leaf size; Flowers or flower buds
Coast live oak ("agrifolia")	Feb 27, 1896	"Some... opening buds... One agrifolia has set of new leaves dark purple + velvety – another has yellow-green lvs"[2]	Breaking leaf buds; Young leaves
Live oak [species unknown]	Mar 30, 1895	"young leaves + shoots + blossom"[1]	Young leaves; Flowers or flower buds
	Mar 8, 1896	"Some of the live oaks just opening buds. Other with shoots six inches long. So also the blue oaks. + White-oaks."[2]	Breaking leaf buds
Oak [species unknown]	Apr 12, 1895	"full leaf"[1]	Leaves
	Feb 16, 1896	"beginning to [illegible] buds"[2]	Breaking leaf buds (absence)?
	Mar 6, 1896	"The oaks rapidly developing buds"[2]	Breaking leaf buds (absence)?
	Mar 17, 1896	"oaks getting their leaves – nearly all now show yellow, even the latest"[2]	Breaking leaf buds? Increasing leaf size?
"Ribes"	Feb 14, 1896	"in flower"[2]	Flowers or flower buds
Valley oak ("Lobata," "White oak")[4]	Mar 30, 1895	"yellow-green + purplish in foliage... leaves... not yet full grown"[1]	Increasing leaf size
	Feb 27, 1896	"Some... opening buds. One... has leaves nearly unfolded must have buds close."[2]	Breaking leaf buds
	Mar 1, 1896	"One... in leaf the largest 1 inch long others not yet opened buds"[2]	Increasing leaf size; Breaking leaf buds (absence)
	Mar 8, 1896	"Some of the live oaks just opening buds. Other with shoots six inches long. So also the blue oaks. + White-oaks."[2]	Breaking leaf buds
	Mar 27, 1896	"The two white-oaks not yet in full leaf. Some just opening buds, but most are well clad with soft downy new-born leaves. Some of their young shoots 5 inches long + fertile flowers open."[2]	Breaking leaf buds; Increasing leaf size; Flowers or flower buds; Open flowers
Willow [species unknown]	Mar 19, 1872	"young leaves"[3]	Increasing leaf size
	Jan 24, 1895	"shedding pollen"[1]	Pollen release

Table 7. Historical observations related to the timing of natural history events for trees and shrubs in and around the study area, and corresponding "phenophases" as described by Denny et al. 2014. Where possible, variations in the original terminology have been combined into a single group or common name in the Historical Observation column (e.g., "Black oak" includes references to "Kellogg oak" and "Cal oak"). The terms in the Phenophase column correspond to standardized phenophases for different plant functional groups as defined in Denny et al. 2014, and in some cases indicate the absence of the specified phenophase.

[1] Muir 1895

[2] Muir 1896

[3] Strentzel 1850-82

[4] Observations of "white oak" assumed to refer to *Quercus lobata*

other groups, such as other grasses, Low-spine Compositae (e.g., *Ambrosia* sp.), Cyperaceae (sedge family), Cruciferae (Brassicaceae; mustard family), *Eriogonum* (buckwheat), *Euphorbia* (spurge), Labiatae (Lamiaceae; mint family), Leguminosae (Fabaceae; bean family), Onagraceae (evening primrose family), Polemoniaceae (phlox family), Rhamnaceae (buckthorn family), *Rhus radicans/integrifolia* (poison ivy/lemonade berry), Rosaceae (rose family), and ferns. The brick also contained pollen of some introduced genera, such as *Phoenix* (date palm) and *Erodium* (filaree).

In addition to detailed descriptions of vegetation and plant species, historical textual sources make references to numerous animal species observed in the vicinity of Mt. Wanda or the Muir ranch. Notably, John Muir and others describe a Great Blue Heron (*Ardea herodias*) rookery that was located in Alhambra Valley just east of Mt. Wanda and may have extended into the study area (Muir 1895, Muir 1896, Wheelock and Horsfall 1912; Fig. 19). On February 13, 1896, for example, Muir wrote, "There are 31 heron nests among the sycamores on Alhambra Crk [Creek], 13 on one tree. All the nest trees are sycamores growing on the bank of the valley creek. The overflow from the regular heronry consists of about a dozen nests on large oaks on the hills to the westward" (Muir 1896). Some of the other animal species described in historical accounts include Song Sparrow (*Melospiza melodia*), Western Screech-Owl (*Megascops kennicottii*), House Wren (*Troglodytes aedon*), "winter wren," quail, robins, larks, "zonotrichias" (likely Golden-crowned Sparrow [*Zonotrichia atricapilla*] or White-crowned Sparrow [*Z. leucophrys*]), "bluejays," "sickle-billed thrush," "Louisiana Tanager" (Western Tanager [*Piranga ludoviciana*]), coyotes, and mountain lions (Strentzel 1850-82, Muir 1895, Muir 1896, Muir 1901-2, Bowles 1902; see Appendix).

Figure 19. Great Blue Heron rookery at Pescadero Marsh. Historical documents describe a Great Blue Heron rookery just east of Mt. Wanda. (courtesy of Brian Washburn; License: https://creativecommons.org/licenses/by/2.0/)

PRELIMINARY DATA SYNTHESIS

The dataset assembled for this reconnaissance study has strengths and weaknesses in terms of its utility in reconstructing past ecological conditions. Early maps are for the most part quite coarse and provide very little ecological information about the historical Mt. Wanda landscape. Landscape photographs show details of the structure and composition of vegetation communities, but are for the most part limited to the portions of the study area adjacent to Alhambra Valley and Franklin Canyon. Textual data – in particular John Muir's late 19th century journal entries – provide very detailed information about species presence and phenology, but lack spatial detail. Historical aerial photos and Wieslander VTM maps are spatially explicit and provide coverage of the entire study area, but were produced relatively late and can be challenging to interpret.

Though the assembled dataset does not permit a complete reconstruction of the historical landscape of Mt. Wanda, we are able to draw some preliminary conclusions about early ecological conditions. Forests and woodlands appear to have occupied much of the northern portion of Mt. Wanda, as well as hillsides within Strentzel Canyon. Nineteenth and early-twentieth century sources document the presence of numerous tree species in the area, including blue, black, valley, and coast live oak; California bay; California buckeye; and California black walnut. Grasslands appear to have occupied much of the central, higher-elevation portions of the study area historically. By the late 19th century, the grasslands were largely dominated by non-native species such as wild oats, and also supported occasional shrubs or oaks. A diverse array of native forbs and wildflowers, documented in John Muir's journals and other textual sources, contributed greatly to the botanical diversity of grasslands and forest understories. Historical evidence for chaparral vegetation is limited, though several early landscape photos appear to show small patches of chaparral on the northern slopes adjacent to the Muir House.

The ecological patterns documented in 19th century sources are generally supported by more spatially explicit and comprehensive sources from the early to mid-20th century, including the 1939 aerial photographs (Fig. 21) and the Wieslander VTM map (Fig. 23)[1]. Though the amount of information provided by 19th century sources is not sufficient to rule out the possibility that substantial shifts in vegetation community distribution occurred between the mid-19th and mid-20th centuries, there is no clear evidence to suggest that major changes had occurred. Consistent with earlier sources, both the aerial photographs and VTM map show the northern and southernmost portions of Mt. Wanda dominated by forests and woodlands, with the central portion of the study area dominated by grassland and sparse oak savanna. However, the forest composition documented in the VTM map, which shows large portions of the study area dominated by valley oak or black oak and only a relatively small area dominated by coast live oak and California bay, is not entirely consistent with early textual accounts, which indicate that blue oak, coast live oak, and California bay were also common tree species in the area historically.

Comparison of historical and contemporary data appears to show the persistence of broad patterns of vegetation community distribution over time. As can be seen in modern aerial imagery (Fig. 22) and modern vegetation mapping (Fig. 24), forests and woodlands still occupy the northern portion of the study area and much of Strentzel Canyon, while grasslands dominate high-elevation areas in the center of the study area. Small patches of chaparral or shrubland exist in the northeastern corner of the study area, in the same general area

---

[1] Researchers at the Information Center for the Environment at UC Davis have crosswalked the original species assemblages recorded on the VTM map to vegetation alliances defined in the Manual of California Vegetation (Sawyer et al. 2009).

as the patch of chaparral visible in early landscape photographs. The vegetation distribution shown in the modern aerial imagery is strikingly similar to the patterns visible in the 1939 aerial photographs, though forest canopy cover appears to have increased somewhat since 1939, with many individual trees showing increased crown spread in the 2009 aerials (Fig. 20).

As with vegetation distribution, the composition of vegetation communities on Mt. Wanda today appears to be broadly consistent with historical evidence. Dominant plant species in the modern vegetation mapping (NPS 2004) include species repeatedly documented in historical sources, such as California bay, California buckeye, wild oats, and blue, valley, and coast live oak. In addition, though tree species cannot be determined from the 1939 aerial photographs, many individual trees can be located in both the 1939 and 2009 aerials, suggesting that major changes in overstory species composition did not occur during this interval (Fig. 20). The species composition depicted in the VTM map, which is somewhat inconsistent with earlier sources, also appears to be at odds with modern vegetation mapping. Many of the areas shown as dominated by valley oak or black oak in the VTM map are classified as Blue Oak Woodland, Coast Live Oak Forest, or California Laurel Forest in the modern vegetation mapping (which uses the first version of the National Vegetation Classification Standards). While it is possible that forest species composition has shifted substantially within the past 70 years, with a decline in the abundance of valley oak and black oak and an increase in the abundance of blue oak, coast live oak, and California bay, this scenario appears unlikely. Given the evidence for abundant blue oak, coast live oak, and California bay cover on Mt. Wanda in early textual data, the persistence of many individual trees visible in both the 1939 and 2009 aerial photographs, and the susceptibility of the VTM methodology to errors in species identification, it appears equally if not more plausible that the VTM surveyors misidentified or omitted several dominant trees species on Mt. Wanda.

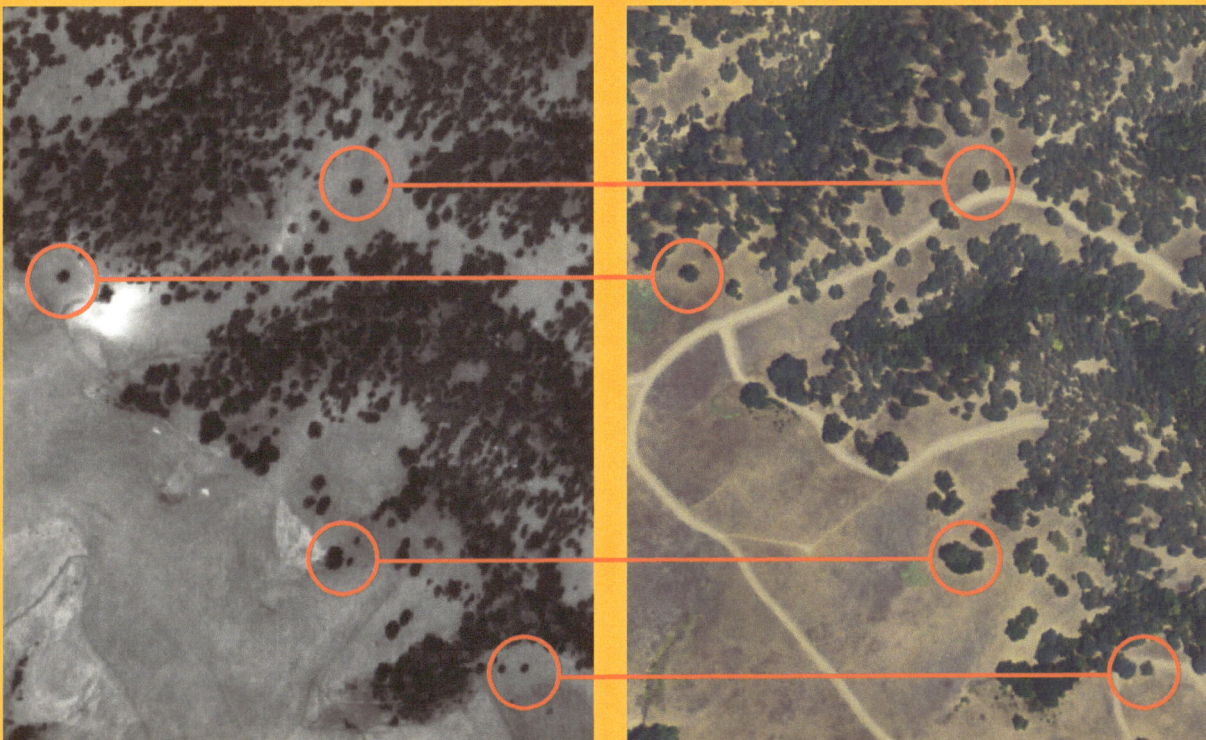

Figure 20. Historical and modern aerial photos highlighting individual trees visible in both images. (USDA 1939, 2009)

Figure 21. The Mt. Wanda landscape visible in the 1939 aerial photographs appears to reflect the same general ecological patterns documented in 19th century sources. Forests and woodlands cover the northern and southernmost portions of the study area, while grasslands and oak savanna occupy the central, higher-elevation portions. (USDA 1939)

Figure 22. The broad patterns of vegetation community distribution documented in the 1939 aerial photographs and other historical sources have persisted to the present day, as can be seen in the 2009 aerial photographs. Forest canopy cover appears to have increased in many areas since 1939. (USDA 2009)

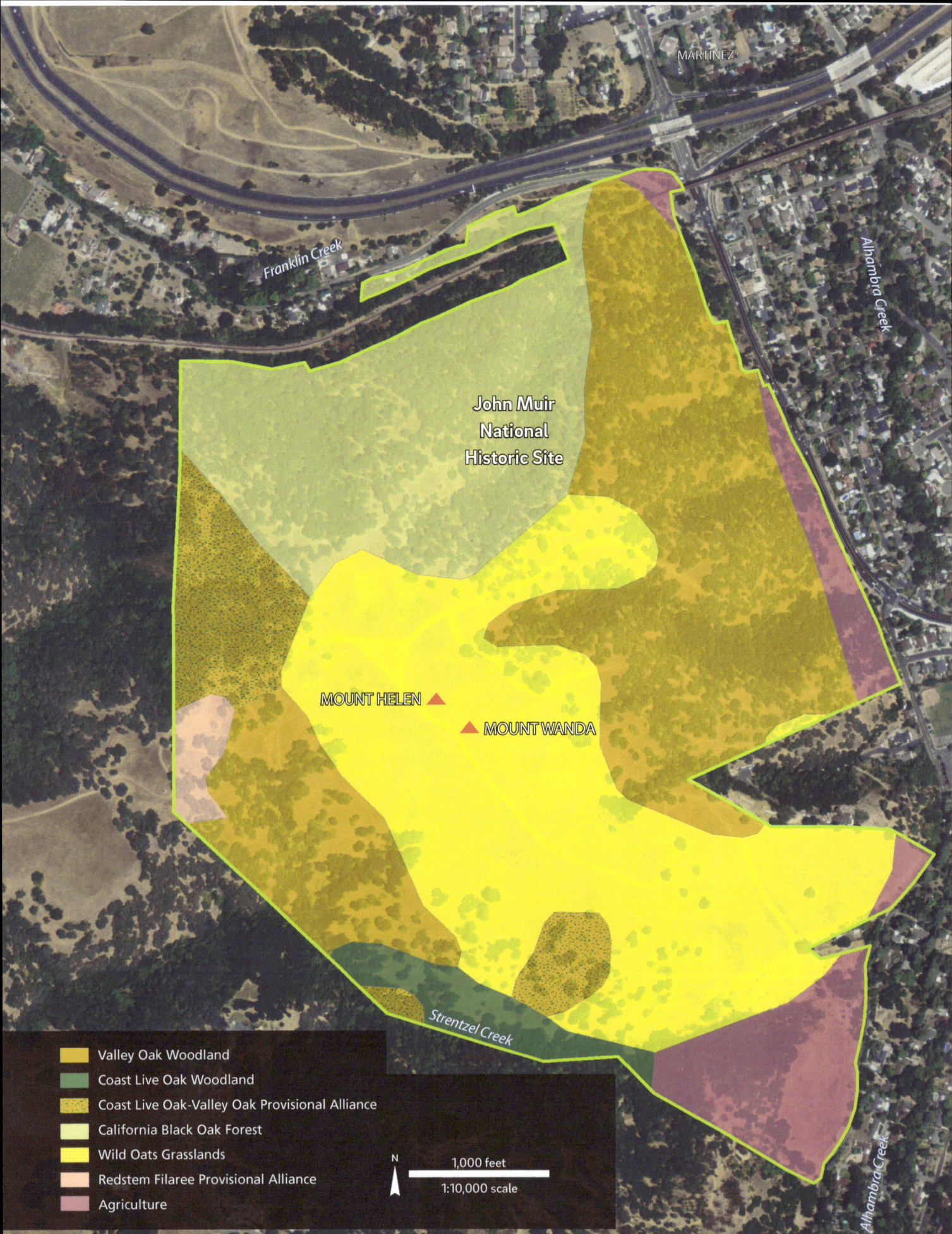

Legend:

- Valley Oak Woodland
- Coast Live Oak Woodland
- Coast Live Oak-Valley Oak Provisional Alliance
- California Black Oak Forest
- Wild Oats Grasslands
- Redstem Filaree Provisional Alliance
- Agriculture

N

1,000 feet

1:10,000 scale

Figure 23. The above map shows the species assemblages recorded on the 1933 Wieslander VTM map of Mt. Wanda (see Fig. 9 on pages 14-15) crosswalked to vegetation alliances defined in the Manual of California Vegetation (Sawyer et al. 2009). Discrepancies between the species recorded by VTM surveyors and both 19th century records and modern vegetation mapping (opposite) may indicate changes in vegetation community composition, but are more likely the result of inaccuracies in the VTM mapping. (courtesy of Information Center for the Environment, UC Davis)

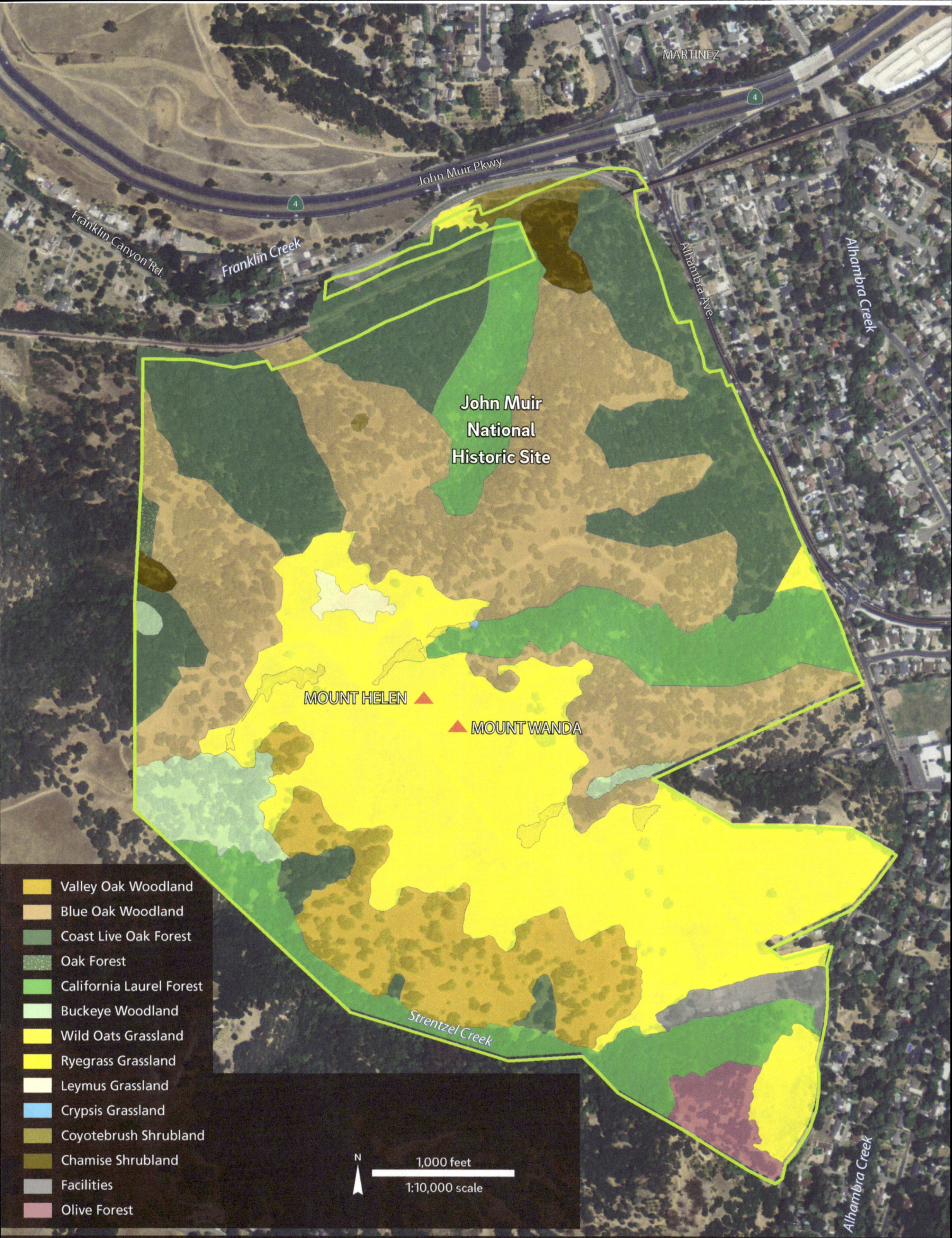

Figure 24. The contemporary vegetation distribution on Mt. Wanda reflects the general patterns documented in historical sources. Blue Oak Woodland, Coast Live Oak Forest, and California Laurel Forest dominate the northern parts of Mt. Wanda. Grasslands occupy the higher-elevation regions in the center of the study area, while Valley Oak Woodland and California Laurel Forest occupy slopes within Strentzel Canyon to the south. (NPS 2004)

Legend

- Valley Oak Woodland
- Blue Oak Woodland
- Coast Live Oak Forest
- Oak Forest
- California Laurel Forest
- Buckeye Woodland
- Wild Oats Grassland
- Ryegrass Grassland
- Leymus Grassland
- Crypsis Grassland
- Coyotebrush Shrubland
- Chamise Shrubland
- Facilities
- Olive Forest

Map labels: MARTINEZ, John Muir Parkway, Franklin Canyon Rd., Franklin Creek, Alhambra Ave., Alhambra Creek, John Muir National Historic Site, MOUNT HELEN, MOUNT WANDA, Strentzel Creek, Alhambra Creek

N — 1,000 feet — 1:10,000 scale

NEXT STEPS

This study is intended as a first step towards reconstructing the historical ecology of Mt. Wanda. The data collection, compilation, and analyses conducted thus far provide a platform for further research and analysis that can deepen and broaden our preliminary findings and fill in gaps in our understanding of the historical Mt. Wanda landscape. The highest-priority next step for expanding on the research presented here is to conduct additional historical data collection and compilation, which will in turn support historical habitat mapping and more thorough landscape change analysis.

Though much of the most easily accessible historical data has already been obtained, it is likely that a large amount of additional relevant material remains to be collected and analyzed. Several libraries and archives already visited for this study are highly likely to have additional relevant holdings, including The Bancroft Library at UC Berkeley, the Contra Costa County Public Library, and the University of the Pacific Holt-Atherton Special Collections (UOP). In particular, the Special Collections at UOP may contain a substantial amount of additional relevant material in the John Muir Papers (MSS 048), Muiriana Collection (MSS 307), and James Eastman Shone Collection of Muiriana (MSS 301). Other archives and institutions not visited for this study may also have relevant holdings, including local archives (e.g., Martinez Historical Society), county archives (e.g., Public Works Department, Resource Conservation District), and regional archives (e.g., California State Archives, California State Library, California State Railroad Museum). Numerous online databases also likely contain a substantial amount of additional relevant data.

Based on the results of this reconnaissance study, it appears that developing a historical habitat map for Mt. Wanda at a scale or resolution comparable to that of modern JOMU vegeta-

tion mapping is probably not possible given the limitations of the available data. However, developing coarser-scale historical habitat mapping, drawing on early records but guided largely by spatially explicit 20th century data sources such as the 1939 aerial photos and the Wieslander VTM map, would likely be feasible. In addition to habitat mapping, other promising areas for future research might include more in-depth assessment of changes in vegetation patterns over time, analysis of the historical drainage network and hydrologic characteristics, documentation of wetland areas within the study area, and further examination of wildlife records and species support functions.

In addition to further archive-based research and analysis, a variety of other research avenues could be taken to complement this work or apply the preliminary historical ecology findings to address scientific and management questions:

- Reconcile the discrepancies between the Wieslander VTM map and the 1939 aerial photos to create a synthetic map of 1930s vegetation communities, which could then be crosswalked to contemporary vegetation mapping to analyze changes in habitat distributions over time.

- Compare historical phenology data with modern records to identify potential changes in the timing of life history events for plant and animal species found on Mt. Wanda.

- Obtain tree cores and conduct tree ring analyses to increase understanding of forest age distribution, recruitment history, and disturbance history.

- Collect and analyze sediment cores from Mt. Wanda to increase understanding of paleoecological conditions and historical vegetation composition.

Figure 25. John Muir near his home in Martinez, CA, circa 1907-9. (left to right: photos #F24-1314, #F24-1307, John Muir Papers, Holt-Atherton Special Collections, University of the Pacific Library. ©1984 Muir-Hanna Trust; photo #1.3.3.3.10, James Eastman Shone Collection of Muiriana, Holt-Atherton Special Collections, University of the Pacific Library. ©1984 Muir-Hanna Trust)

ACKNOWLEDGMENTS

We would like to thank Fernando Villalba at the National Park Service (NPS) for helping to develop the vision for the project and for his support and guidance throughout the course of the study. Thanks to Sara Hay at NPS for providing access to the JOMU archival collections and assistance with data collection. We are indebted to all of the staff and volunteers at the archives we visited throughout the course of the project, including The Bancroft Library, Bureau of Land Management, Contra Costa County Historical Society, UC Berkeley Earth Science & Map Library, John Muir National Historic Site, and University of the Pacific Holt-Atherton Special Collections. Thank you to Jackie Bjorkman and Jim Thorne at the Information Center for the Environment (ICE) at UC Davis for providing access to Wieslander VTM data and assistance with interpretation. We owe a special thanks to Barbara Postel, who enthusiastically volunteered her time to assist with archival research and contributed to the collection of a robust dataset. We are grateful to the Friends of Alhambra Creek for their support of the project and recommendations for potential source institutions. Two interns provided valuable research assistance: Rachel Powell, from the Bill Lane Center for the American West at Stanford, and Jessica Sanchez, from NPS. Additional research assistance was provided by SFEI staff members Sam Safran, Carolyn Doehring, and Micha Salomon. This study was funded by the National Park Service.

Near summit of Mt. Wanda. (photograph by SFEI, July 2013)

REFERENCES

Bowles JH. 1902. The Louisiana Tanager (*Piranga ludoviciana*). *The Condor* 4(1):16-17.

Brown EC. 1891. Plat of a portion of the Alhambra Ranch, Contra Costa Co, Cal. JOMU 3889, John Muir National Historic Site. *Courtesy of John Muir National Historic Site.*

Carpenter EJ, Cosby SW. 1933. Soil map: Contra Costa County, California. U.S. Department of Agriculture, Bureau of Chemistry and Soils. Series 1933.

Carpenter EJ, Cosby SW. 1939. *Soil Survey of Contra Costa County, California.* U.S. Department of Agriculture. Bureau of Soils. Series 1933. Washington, DC: Government Printing Office.

County of Contra Costa. 1888. Mortgage document between O.C. Huefner and John Strentzel, 11 May 1888 (filed 14 May 1888), Contra Costa County, California, Mortgages Vol. 24, page 590. John Muir Papers, MSS 048, Related Papers of William F. Badè, Series Va, Box 8: Clippings on Muir Selected by Badè, 1915-1934. Holt-Atherton Special Collections, University of the Pacific Library.

County of Contra Costa. 1892. Deed of Sale from John Muir to Roger Cutler, 25 July 1892, Contra Costa County, California, Deed Book 62, page 334. County Clerk-Recorder's Office, Martinez, CA.

County of Contra Costa. 1893. Deed of Sale from John Muir to Roger Cutler, 22 December 1892 (filed 3 April 1893), Contra Costa County, California, Deed Book 63, page 410. County Clerk-Recorder's Office, Martinez, CA.

Cummings LS, Puseman K. 1991. Pollen, phytolith, and macrofloral analysis at the John Muir House, California. PaleoResearch Laboratories. Denver, CO. doi:10.6067/XCV8KP8199. http://core.tdar.org/document/375335. Accessed June 11, 2014.

Cushing JK. 1921. John Muir's stately home to be sold. One of the nation's landmarks to pass. *San Francisco Examiner.* John Muir Papers, MSS 048, Related Papers of William F. Badè, Series Va, Box 8: Clippings on Muir Selected by Badè, 1915-1934. Folder 69.16: Newspaper Clippings, 1921. Holt-Atherton Special Collections, University of the Pacific Library.

de Anza JB, Bolton HE. 1930. *Anza's California expeditions.* Berkeley, CA: University of California Press.

Denny EG, Gerst KL, Miller-Rusing AJ, et al. 2014. Standardized phenology monitoring methods to track plant and animal activity for science and resource management applications [electronic supplementary material: Plant and animal phenophase definitions]. *International Journal of Biometeorology* 58:591-601. https://www.usanpn.org/files/shared/files/Plant%20and%20Animal%20Phenophase%20Definition%20Supplement.pdf. Accessed June 11, 2014.

D'Antonio CM, Malmstrom C, Reynolds SA, Gerlach J. 2007. Ecology of invasive non-native species in California grassland. In *California grasslands: Ecology and management*, ed. Mark R. Stromberg, Jeffrey D. Corbin, and Carla M. D'Antonio, 67-83. Berkeley, CA: University of California Press.

DiTomaso JM, Kyser GB, et al. 2013. *Weed control in natural areas in the western United States.* Weed Research and Information Center, University of California. 544 pp. http://wric.ucdavis.edu/information/natural%20areas/wr_A/Avena_barbata-fatua.pdf. Accessed February 20, 2015.

Engelmann G. George Engelmann to John Muir, 11 April 1881. John Muir Papers, MSS 048, Correspondence & Related Documents, Series I. Holt-Atherton Special Collections, University of the Pacific Library. http://digitalcollections.pacific.edu/cdm/compoundobject/collection/muirletters/id/10481/rec/2. Accessed June 10, 2014.

Fages P, Treutlein TE. 1972. Fages as Explorer, 1769-1772. *California Historical Society* 51(4):338-356.

Guma IR, Pérez de la Vega M, García P. 2006. Isozyme variation and genetic structure of populations of *Avena barbata* from Argentina. Genetic Resources and Crop Evolution 53:587-601.

Hays RB. 1852. *Field notes of the survey of the township lines north of the base line, west of the meridian and south and east of Suisun, San Pablo and San Francisco Bay, as made by Robert B. Hays under his contract of 16 February 1852.* U.S. Department of the Interior, Bureau of Land Management. Book 210-2. *Courtesy of Bureau of Land Management.*

Hunter JC, Veirs SD, Reeberg PE. 1993. *The flora, vegetation, and human use of Mt. Wanda, John Muir National Historic Site, Martinez, California.* Cooperative Park Studies Unit, National Park Service, Davis, California.

Jepsen EPB, Murdock AG. 2002. *Inventory of native and non-native vegetation on John Muir National Historic Site, Eugene O'Neill National Historic Site, and Port Chicago National Monument.* National Park Service report (in cooperation with Point Reyes Bird Observatory).

Keeler-Wolf T. 2007. The history of vegetation classification and mapping in California, Chapter 1. In *Terrestrial vegetation of California, 3rd edition*, ed. MG Barbour, T Keeler-Wolf, and AA Schoenherr, 1-42. Berkeley, Los Angeles, London: University of California Press.

Keibel JA. 1999. *The Alhambra Valley trestle. Then and now: a centennial.* Concord, CA: J.A. Keibel.

Kelly M, Allen-Diaz B, Kobzina N. 2005. Digitization of a historic dataset: the Wieslander California Vegetation Type Mapping project. *Madroño* 52(3):191-201.

Killion J. 2005. *Cultural landscape report for John Muir National Historic Site. Volume 1: Site history, existing conditions, and analysis.* Olmsted Center for Landscape Preservation. National Park Service, Boston, Massachusetts.

Martinez Y. 1837. U.S. v. María Antonia Martinez de Richardson et al., Land Case No. 205 ND [Pinole], docket 334, U.S. District Court, Northern District. *Courtesy of The Bancroft Library, UC Berkeley.*

Milliken R. 1995. *A time of little choice: the disintegration of tribal culture in the San Francisco Bay Area 1769-1810.* Menlo Park, CA: Ballena Press.

Moore C. 2006. *Watershed management report. John Muir National Historic Site, Martinez, California.* Natural Resource Technical Report NPS/SFANNRTR—2006/022. National Park Service, Fort Collins, Colorado.

Muir H. Helen Muir to John Muir, 10 July 1896. John Muir Papers, MSS 048, Correspondence & Related Documents, Series I. Holt-Atherton Special Collections, University of the Pacific Library. http://digitalcollections. pacific.edu/cdm/compoundobject/collection/muirletters/id/1879/rec/4. Accessed June 11, 2014.

Muir H. 1901-2. [Helen Muir diary]. *Courtesy of John Muir National Historic Site.*

Muir J. 1895. *49. January-July 1895, ranch life (Martinez, California).* John Muir Papers, MSS 048, Journals & Sketchbooks, Series II. Holt-Atherton Special Collections, University of the Pacific Library. http://digitalcollections.pacific.edu/cdm/compoundobject/collection/muirjournals/id/2759/rec/1. Accessed June 10, 2014.

Muir J. 1896. *51. January-May 1896, ranch life (Martinez, California).* John Muir Papers, MSS 048, Journals & Sketchbooks, Series II. Holt-Atherton Special Collections, University of the Pacific Library. http://digitalcollections.pacific.edu/cdm/compoundobject/collection/muirjournals/id/2842/rec/2. Accessed June 10, 2014.

NPS (National Park Service). 1991. *General management plan and environmental assessment. John Muir National Historic Site, California.* Western Regional Office, National Park Service.

NPS (National Park Service). 2004. Vegetation map – Mt. Wanda, John Muir National Historic Site – 2004. Inventory and Monitoring Program, National Park Service. San Francisco, CA. *Courtesy of John Muir National Historic Site.*

NPS (National Park Service). 2014. Species list [John Muir National Historic Site, vascular plants]. NPSpecies. https://irma.nps.gov/NPSpecies/. Accessed June 11, 2014.

O'Neil S, Egan S. 2004. *Plant community classification and mapping project: John Muir National Historic Site (Mt. Wanda)*. National Park Service.

Pacific Rural Press. 1889. Alhambra Valley. January 12. *Courtesy of California Digital Newspaper Collection.*

Sargent, CS. C[harles] S[prague] Sargent to John Muir, 19 February 1908. John Muir Papers, MSS 048, Correspondence & Related Documents, Series I. Holt-Atherton Special Collections, University of the Pacific Library. http://digitalcollections.pacific.edu/cdm/compoundobject/collection/muirletters/id/5228/rec/1. Accessed June 10, 2014.

Sawyer JO, Keeler-Wolf T, Evens J. 2009. *A manual of California vegetation, second edition*. Sacramento, CA: California Native Plant Society.

Soil Survey Staff. 2014. Web Soil Survey. U.S. Department of Agriculture, Natural Resources Conservation Service. http://websoilsurvey.sc.egov.usda.gov/App/HomePage.htm. Accessed March 13, 2015.

Stanford B, Grossinger RM, Askevold RA, et al. 2011. *East Contra Costa County historical ecology study*. Prepared for Contra Costa County and the Contra Costa Watershed Forum. A report of SFEI's Historical Ecology program, SFEI Publication # 648. San Francisco Estuary Institute, Oakland, CA.

Strentzel JT. n.d. Biography of John Theophil Strentzel. Strentzel family papers, BANC MSS 75/86 c, The Bancroft Library, University of California, Berkeley.

Strentzel LE. 1850-82. Louisiana Erwin Strentzel papers, 1850-1882. BANC MSS C-F 16, The Bancroft Library, University of California, Berkeley.

Taylor KW. 1865. Plat of the Pinole Rancho finally confirmed to M.A. M. de Richardson. U.S. Surveyor General's Office. San Francisco, CA. *Courtesy of Bureau of Land Management.*

USDA (U.S. Department of Agriculture, Western Division Laboratory). 1939. [Aerial photos of Contra Costa County]. Scale: 1:20,000. Agricultural Adjustment Administration (AAA). *Courtesy of Earth Sciences & Map Library, UC Berkeley.*

USDA (U.S. Department of Agriculture). 2009. [Natural color aerial photos of Contra Costa, Sacramento, San Joaquin, Solano, Yolo counties]. Ground resolution: 1m. National Agriculture Imagery Program (NAIP). Washington, D.C.

USGS (U.S. Geological Survey). [1893-4]1897. Concord Quadrangle, California: 15-Minute series (Topographic). 1:62,500.

USGS (U.S. Geological Survey). [1893-4]1915. Concord Quadrangle, California: 15-Minute series (Topographic). 1:62,500.

Wagner T, Sandow G. 1894. Map showing portions of Alameda and Contra Costa Counties, City and County of San Francisco, California: carefully compiled from official and private maps, surveys and data. *Courtesy of Contra Costa County Public Works Department.*

Wheelock IG, Horsfall RB. 1912. *Birds of California: an introduction to more than three hundred common birds of the state and adjacent islands with a supplementary list of rare migrants, accidental visitants, and hypothetical subspecies*. Chicago: A. C. McClurg & Co.

Wieslander AE, Yates HS, Jensen HA, Johannsen PL. n.d. *Manual of field instructions for vegetation type map of California*. Unpublished report. http://digitalassets.lib.berkeley.edu/vtm/ucb/text/cubio_vtm_fm.pdf. Accessed June 10, 2014.

APPENDIX: SELECTED QUOTES

This appendix includes selected quotes or excerpts from primary sources that contain relevant information about the historical ecological or physical environment of Mt. Wanda or surrounding areas. Quotes are organized according to topic and then chronologically; where a single quote relates to multiple topics, it was included only in the single most relevant section.

Vegetation

General

1853: "Here was a lovely fertile valley, protected by high hills, from the cold winds and foggs [sic] of San Francisco, a stream of living water flowing through it, the hills and valley partially covered with magnificent laurel, live- oak and white- oak trees, and everywhere a green mantle of wild oats from 8- 12 inches high." (Strentzel n.d.)

1868, Apr 5: "This is just such a morning as that was fifteen years ago… I could almost see my little children trying to wade through the wild oats which were at that time over one foot high, and wet with a heavy dew. The hills present the same appearance they did then, except that the trees have grown larger. Vegetation is very around on the hills [sic] it was difficult for me to realize that fifteen years had passed away, so little have they changed in that time. But how different in the valley. When we first came here, a great portion of it was entirely covered with chaparral, blackberry briars, alder, willows &c [sic] now it is smiling with lovely vines and fruit trees all in bearing." (Strentzel 1850-82)

1870, May 9: "Hills begin turn [sic] yellow." (Strentzel 1850-82)

1872, Mar 19: "The sun cam shining down aslant the hills, throwing a flood of light over the pale green slopes, and the dark rich green of the buckeyes, laurels and live oaks while the young leaves of the sycamore and willows glittered in the sun like diamonds." (Strentzel 1850-82)

1878, Sept 3: "Drove up to L. Smith's… Had a delightful ride home. This valley and cañon is [sic] always gloriously beautiful, but never more so to me than now, the rich green foliage of the tree contrasting finely with the brown hills, the live oak trees along the creek, particularly the one this side the old Holliday house which the Dr. has always loved so, and once said when we were passing it 25 years ago, that he would give any price if it was growing in our yard. It is now a grand large tree and covered with dense masses of rich foliage." (Strentzel 1850-82)

1895, Jan 24: "Laurel in full flower. Dodecatheon in bud. Buttercups in flower. Willows shedding pollen. Nemophila, chickweed mustard [sb: preceeding 2 words difficult to read] in flower. The latter 4 ft high + gone to seed some of it." (Muir 1895)

1895, Apr 12: "Took a fine fragrant walk up the West hills with Wanda + Helen who I am glad to see love walking – flowers – trees + every bird + beast + creeping thing – Buttercups, clovers, gilias, Brodiaeas, Aliums, Dodecatheons, larkspurs, and portulaccas are in flower. The oaks are in full leaf." (Muir 1895)

1895, May 14: "Hills beginning to show tawny." (Muir 1895)

1895, June 13: "Hills fairly brown now." (Muir 1895)

1896, Feb 3: "Clear many plants in flr." (Muir 1896)

1896, Feb 14: "Took my first ride up the Valley since my sickness. Also walked in the PM with Helen. Nuttalia, Ribes, Laurel, buttercup, in flower… Bees and butterflies abundant." (Muir 1896)

1896, Feb 16: "Walked on the hills with Helen, up Wanda + down the Helen hill. The buckeye will soon be in full leaf. A few of the oaks also beginning to [illegible] buds. Soap-roots + Castilleja in flower." (Muir 1896)

1896, May 15: "Hills still green + flowery." (Muir 1896)

1896, May 24: "Rode up the Valley with L. + the children. Vegetation still fresh on account of rain + coolness. Only the very driest hills show brown as yet. Usually all are tawny ere this." (Muir 1896)

1896, July 10: "The hills are very dry now and there are hardly any flowers on them. Nearly all the apricots are gone." (Muir 1896, Letter from Helen Muir to John Muir)

1921: "The wooded hills that climb up almost from the back porch of the home." (Cushing 1921)

Trees

1881, Apr 11: "As I certainly shall not (if at all) be there in season for the flowering of the oaks, I would ask you to get for me at least the two species which grow close about you. One is Quercus lobata, which grows as you know, above the stables on the Creek; the other is Quercus Douglasii, the blue mountain oak, on the bare hills above you. A few specimens of each of them in flower and again a week or two later when the leaves are not yet fully grown (showing the female flowers or young acorns) would be very desirable. I would say that I found a few specimens of the lobata also on the hills, but Douglasii is readily distinguished by the smoother, whiter bark, the smaller and less lobar leaves; it seems to be the common tree on the arid hills." (Engelmann 1881)

1889: "Prof. John Swett, four miles from Martinez. Six years ago the professor began to improve this portion of the valley, and now has a country seat whose loveliness can be realized only by personal inspection. The valley here is about one mile wide and surrounded with hills covered with evergreen trees, except in portions where the oaks have been leveled and their places taken by the choicest vines, deciduous fruits and olive trees." (Pacific Rural Press 1889)

1895, Jan 26: "Leaves opening rapidly on the buckeyes – even the buds make a fine show, so large they are + so fine a purple tinge makes petal-like the scales." (Muir 1895)

1895, March 30: "The foliage of the woods on the hills now charmingly fresh + beautiful + the grass green. This scenery of our valley at its best. The Kellogg Oak with yellow-green foliage is now nearly in prime. The first to put forth its leaves. The Douglas + lobata also yellow-green + purplish in foliage but less lively in color + later in putting forth leaves – they are not yet full grown. The live oak is brown with its young leaves + shoots + blossom. The pea green foliage of buckeye has fine effect. The first of all to appear." (Muir 1895)

1896, Feb 18: "Had walk with the children up the hills opposite the Heronry to branch overflow Heronry on large oak with only 3 nests -- + one wh has been barely commenced. The Kellogg or Cal Oak is now in full flower. The pollen ripe, some catkins 4 inches long + [next phrase illegible]. Some leaves 1 ½ inch long. Purplish red at first. Dodecatheon [in?] flr. Blue oak buds not yet open." (Muir 1896)

1896, Feb 24: "A week to ten days difference in time of flowering of Kellogg Oak. Some lvs two inches long. Dodecatheon + Saxifrage in flr." (Muir 1896)

1896, Feb 27: "Some of the live + white oaks opening buds. One white oak has leaves nearly unfolded must have buds close. One agrifolia has set of new leaves dark purple + velvety – another has yellow-green lvs. Most stuff in bud." (Muir 1896)

1896, March 1: "Helen + I made an excursion to the big Laurel. It has spread of about 150 ft. Has six trunks growing from old burned out stump, the largest about 6 ft dia. These main trunks are being reinforced by six smaller ones. [illegible] in growth – perhaps immortal. Young trees [coming?] on Live oak shoots 5 inches long grown in one week. One of the white-oaks in leaf the largest 1 inch long others not yet opened buds. The rate of growth of the young shoots is wonderful." (Muir 1896)

1896, March 6: "The oaks rapidly developing buds." (Muir 1896)

1896, March 8: "In the evening Helen + I took a walk in the West hills. Some of the live oaks just opening buds. Other with shoots six inches long. So also the blue oaks. + White-oaks. The different tones of brown + purple + yellow green swelling in rounded broad masses on the hills very fine." (Muir 1896)

1896, March 12: "The oaks with varying tones of yellow brown + green making delightful pictures on the hills. A few of the blue oaks not yet in leaf – buds just beginning to open, a few are open + male flrs in full fringe." (Muir 1896)

1896, March 17: "Grass growing fast + the oaks getting their leaves – nearly all now show yellow, even the latest." (Muir 1896)

1896, March 27: "Took long walks over the west hills + far [away?] with Helen. The two white-oaks not yet in full leaf. Some just opening buds, but most are well clad with soft downy new-born leaves. Some of their young shoots 5 inches long + fertile flowers open. The young buds for next yrs leaves a fine rose color." (Muir 1896)

1908, Feb 19: "We have here In the herbarium very unsatisfactory material of the California Walnut which, as you Know, does not grow very far from your house; at least we drove to it once in a comparatively short time. I am rather interested in the matter, for it seems possible that the little bush in southern California with small nuts may be a different species from your tree. Wouldn't it be possible for you to press for me some good flowering specimens this spring when the trees are in bloom and later make some good leaf specimens?" (Sargent 1908)

Grasslands

1881 Feb 19: "Hills covered with grass several inches high, and the almond trees in full bloom." (Strentzel in Killion 2005)

1895, July 18: "Walked with Helen up the West Hill above the orchard. The Sierra smothered in dust + smoke + invisible. The pastures are all overrun with Chili Thistle. Scarce any grasses left. Now dry + dead. Gnaphalium, a woody [illegible] hosackia, Monardella, bush mimulus still in flower. Comparatively few gophers in the hills." (Muir 1895)

1896, May 8: "Walked over the hills with Wanda + Helen. How the wind did surge... + how the wild oats danced + rippled + clapped their spikelets like happy hands in a passion of joy – (like wild oats on the brow of a hill in a windy day) with back to wind the spikelets showed sooty black." (Muir 1896)

Forbs/Wildflowers

1895, Apr 28: "Made a good long excursion with Helen + Wanda on the West Hills above [illegible]. The flrs mostly gone to seed. Found one nemophila 'baby-blue-eye'. Nearly extinct hereabouts – once abundant – so also most of the gilias. Found one large patch of orthocarpus purple – also white + yellow." (Muir 1895)

1902 Feb 27: "Along the hard sandy road we went as far as the viaduct, here we turned to the right and went up the bank. At the top I picked a little bunch of butter-cups, then we walked through the tunnel... On the way back I found some lovely Nuttalia blooming under some bushes close to the track on the south side of the first cut." (Muir 1901-2)

1896, March 27: "Noticed delicate Collinsia sparsiflora in flr." (Muir 1896)

1896, March 31: "Walked yesterday on hills with Helen. Found tiny buttercup. + Valerianella – a curious little plant." (Muir 1896)

1896, Apr 2: "Walk on the two hills with Helen – a 2nd species of Wythea near summit of Helen hill." (Muir 1896)

Animals

Birds

1850, Aug 20: "This evening Mr. White killed a lot of quail which was a fine stew with potatoes." (Strentzel 1850-82)

1870, Sept 6: "Hardy fix pump killed five bluejays again." (Strentzel 1850-82)

1870, Oct 3: "L. killed 9 quails at one shot!" (Strentzel 1850-82)

1871, Jan 15: "Went Dr. to the Frazer place walked to the top of the hill and all around, saw thousands ducks and snipes." (Strentzel 1850-82)

1895, Jan 23: "Robins are flying in large flocks, driven down from the hills + mtns by the storm. Frogs are singing lustily, + there is always a crowd of handsome zonotrichias about the house hedge. The cats capture many of them." (Muir 1895)

1895, Jan 25: "Saw a song sparrow on the ground in old cherry orchard. It pushed its way through wet Portulacca, nibbling seeds." (Muir 1895)

1895, Jan 26: "Larks – saw a large flock on our alfalfa field 100 or so." (Muir 1895)

1895, Jan 29: "I caught a small screech owl (Scops asio) this morning – a handsome horned fellow with gold eyes." (Muir 1895)

1895, Jan 29: "Larks in flock of 50 seen in Lower Vineyard." (Muir 1895)

1895, Feb 11: "Flocks of robins came today as usual with the storm." (Muir 1895)

1895, Feb 12: "Robins flying about in large restless flocks." (Muir 1895)

1895, Feb 15: "The Blue Herons have come. Mr. [Morey?] Coleman syas they arrived several days ago. They are wonderfully regular in timing their year's affairs. Their clock does not even seem to be affected by the weather." (Muir 1895)

1895, Feb 17: "The herons came last Thursday according to my brother." (Muir 1895)

1895, Feb 18: "Large flocks of robins." (Muir 1895)

1895, Feb 20: "Saw a beautiful song sparrow on the bank of the creek singing cheerily." (Muir 1895)

1895, March 17: "The herons busy egg-laying." (Muir 1895)

1895, Apr 14: "The Louisiana Tanager has been here a week." (Muir 1895)

1895, Apr 22: "The hill vegetation has mostly gone to seed + leaves are fading. A couple of Parkmann's house wrens are building a nest in the woodpile, merry singers + sprightly workers, bits of bright unclouded health." (Muir 1895)

1895, Apr 30: "A pair of wrens building nest in woodpile the second this year so far." (Muir 1895)

1895, May 22: "Birds feeding young. Wrens in woodpile have been out of the shell a week." (Muir 1895)

1895, May 26: "Herons busy with their young – saw 3 in one nest." (Muir 1895)

1895, May 30: "The little speckled wrens are now very busy feeding their young. They must carry several hundreds of insects to the nest every day. They can chatter + complain with a large beetle or fly in their bills. This is the Parkmann's House wren, a sweet cheery singer + vigorous scolder of small dogs + cats." (Muir 1895)

1895, June 3: "The wrens seems to have led off their young today or yesterday." (Muir 1895)

1895, June 10: "The Parkmann's wrens have vanished from the woodpile. I suppose the young were able to fly – for weeks the parents supplied them with worms + insects big + small in [illegible] abundance." (Muir 1895)

1895, June 23: "Saw a Carpodacus + Chrysomitris sitting together on telephone wire, almost touching each other." (Muir 1895)

1895, July 4: "The Herons are all on the wing – a few still lingering on the old nests. The wrens are feeding a second brood of young in the old nest." (Muir 1895)

1895, July 8: "The young wrens still in their woodpile nest. The number of insects they eat in a single day is immense. One every 3 or 4 minutes." (Muir 1895)

1895, July 11: "The wrens very busy feeding their growing young in the woodpile. They bring a worm beetle or fly about once a minute to two minutes. The little babies will soon fly." (Muir 1895)

1895, July 12: "The herons still linger about their nests on the sycamores beside our vineyard though the young have all tried their wings a week or more ago." (Muir 1895)

1895, July 16: "The wrens I think are out of the nest trying their wings + their eyes in this fine world." (Muir 1895)

1896, Jan 1: "Larks in large flocks on the meadow. A few Cow-buntings. Zonotrichia common [illegible] in the hedges." (Muir 1896)

1896, Jan 7: "A small hawk sat on the tail of the windmill today while the wheel revolved freely without being disturbed by the motion or by the squawking axle wh required grease." (Muir 1896)

1896, Jan 29: "Saw small hawk on vane of windmill." (Muir 1896)

1896, Feb 3: "John Reid says one of the herons of the nesting village of the sycamores in the vineyard came today + took a general survey of the premises." (Muir 1896)

1896, Feb 5: "Several more herons came looking at their old homes." (Muir 1896)

1896, Feb 8: "The larks busy in the vineyards. The Herons not yet returned to their nests. The early flrs already gone to seed." (Muir 1896)

1896, Feb 9: "Herons beginning to assemble on their old nests. Perhaps 20 have arrived." (Muir 1896)

1896, Feb 13: "There are 31 heron nests among the sycamores on Alhambra Crk, 13 on one tree. All the nest trees are sycamores growing on the bank of the valley creek. The overflow from the regular heronry consists of about a dozen nests on large oaks on the hills to the westward." (Muir 1896)

1896, Feb 29: "We took a walk to the Heronry. Saw about 30, many already sitting. The wind very high threatening to blow nests away out of the tall sycamores wh are just opening lvs + hanging forth their strings of brown heads of flrs. Leaves some of them [sic] nearly an inch long, very hairy when young, brown, velvety, male heads of flrs green, [illegible] red wh brown." (Muir 1896)

1896, March 28: "Walked with Helen to the old ranch watching the herons at the heronry by the way, they are now hatching." (Muir 1896)

1901 Dec 31: "Papa and I had a nice little walk, we went up through the vineyard, saw a dear little winter wren in a pile of grapes." (Muir 1901-2)

1902 Jan 14: "A little Oregon Robin in the Cherkee [sic] rose bush sounded sleeping." (Muir 1901-2)

1902 Feb 18: "I went out on the east porch… larks which were singing gloriously…/ there under the date palm, a sickle-billed thrush." (Muir 1901-2)

1902 Feb 19: "This has been a lovely day, when Papa were [sic] coming from the [illegible] this evening, it was so warm I had to take off my jacket. There were only 11 herons there this evening, but Papa said that when he was up there about a week ago there were fifteen." (Muir 1901-2)

1902 Feb 21: "Under the peack [peach] tree, so about half a dozen Zonatrickias sat on the limbs and looked down… up ran a crested sickle-billed thrush…/ Then there was the sweetest little wren." (Muir 1901-2)

1902 Feb 22: "The birds were rejoyousing [sic]. The larks and the little Carpodcieses were singing." (Muir 1901-2)

1912: "The Great Blue Heron is a common species throughout California, and nests in almost every locality where it is found. At Muir Station, California, there is a large heronry in sycamore trees on the property of Mr. John Muir, and the noise of the young birds at feeding time can be heard half a mile away. The birds return to their heronry in February, and the young are hatched in April, though fresh eggs have been found as late as June 1." (Wheelock and Horsfall 1912)

Mammals

1870, Apr 10: "I went up the cannon [sic] where Dr. killed a lion 16 years ago." (Strentzel 1850-82)

1870, May 4: "A wolf howled up the valley." (Strentzel 1850-82)

1870, June 22: "Rode up the vally [sic] beyond the willow. Stop in shade of old laurel tree. Squirrels very bad." (Strentzel 1850-82)

1870, Oct 17: "L. has killed in all 20 rabbits, 20 squirrels a dozen quails 15 bluejays." (Strentzel 1850-82)

1872, Dec 20: "Today three coyotes tried to catch sheep on hill." (Strentzel 1850-82)

1877, Aug 15: "Coyotes, coons, wildcats &c [sic] are numerous." (Strentzel 1850-82)

1895, Jan 27: "The little hairy terrier pup caught a hare in the vineyard." (Muir 1895)

1895, May 2: "Gophers in destructive abundance eating off entire tops of vines. I suppose on account of the absence of weeds especially clover + alfileria of which they are fond. Also the three species of mustard." (Muir 1895)

1896, July 10: "We, that is Wanda, Enid Bird, and I, went over to Uncle David's house and then we climbed the east hill so we had a delightful walk and we heard a coyote and saw a lot of squirrels." (Muir 1896, Letter from Helen Muir to John Muir)

Reptiles

1896, March 27: "Saw + caught small gray lizard that was shedding its skin – the old skin gray – the new nearly black. First shed about eyes, giving curious look." (Muir 1896)

1896, July 10: "Fourth of July we did not ride to Martinez but staid at home and we girls climbed the big Oak by the fence and we took our lunch in a basket and ate it under a liveoak tree and we saw a big snake with black and white rings and we thought we had better run down to the house." (Muir 1896, Letter from Helen Muir to John Muir)

Streams

1850, July 1: "The river full." (Strentzel 1850-82)

1850, Sept 26: "Mr. Edwards came down to build on his place... they hope the river will fall in a short time so they can work again – their dam is not much injured by the river but several others nearby seemed about entirely washed away." (Strentzel 1850-82)

1871, Jan 8: "Took a walk with Dr. up the cannon [sic] in the pasture to look for water for the cattle. Sun shone oppressively hot, and it was distressing to see the creek all dried up except a few muddy holes which were dug out barrels sunk for them." (Strentzel 1850-82)

1871, July 10: "Early this morning I took a walk up the creek to see about the water, the bed of the creek is dry and dusty, not a drop of water except the well and two barrels sunk in the spring... Seven years ago there was a fire along the creek which burned all the young trees, but a new set have grown up." (Strentzel 1850-82)

1871, Dec 19: "The water comes rushing down the hills into the orchard in torrents and the creek is nearly to the top of the banks." (Strentzel 1850-82)

1871, Dec 23: "The water came rushing down the cañons in several new places. About noon the creek was full and ran over by the graves... The water stood around the China house in a perfect lake, and ran almost a river over the orchard. We never had such a flood before." (Strentzel 1850-82)

1895, Jan 21: "The storm is now in full bloom, scary rain + wind. The stream from Franklin canon [sic] is again over its banks + spreading in a stiff current carrying sand over the farms + cherry orchard. Almost as high as in the grand flood a week or two ago. Never since the orchard has been planted has this stream been as high. 4 inches of sandy mud has on an average been spread over several acres." (Muir 1895)

1896, Jan 27: "The creek over its banks for the first time this year." (Muir 1896)

Climate

1868, Jan 10: "The ground was covered with snow... This is the third snow we have had in this valley during the time we have lived here, nearly fifteen years. The first snow was in the winter of 1859, and fell to the depth of six inches." (Strentzel 1850-82)

1872, Jan 24: "The most fearful gale of wind we ever had at this place, five or more trees have been blown down in sight on the hills." (Strentzel 1850-82)

1896, Jan 29: "Nearly 10 inches of rain has fallen in nearly one continuous storm." (Muir 1896)

1896, March 3: "Hills all white with snow + hail... Of the lower hills it is curious to note how the north slopes are white down to their bases near sea-level while warmer slopes were bare up to 400 ft. Not because of sunshine on them but because of stored heat in the ground." (Muir 1896)

Other

1837: "As of this date, he [Ygnacio Martinez] finds himself possessed of more than three thousand head of cattle; four hundred head of horses, and six hundred head of sheep, and about eighty tame horses, together with more than three hundred head of milk cattle... The greater part of it [Rancho El Pinole] is indeed unfit for the pasturage of cattle by reason of the stony hills and esteros that are found upon it; and the best part lies in the direction of the 'Sisca' and the 'Cañada del Hambre,' which is most frequented by the stock, and is notorious to all the neighboring inhabitants." (Martinez 1837)

1870, July 9: "Fire in a tree on the hill." (Strentzel 1850-82)

1878, Jan 23: "The cars have not been running since the rains, great quantities of earth slides down from the hills." (Strentzel 1850-82)

1878, Apr 17: "It is fearful to see what great slides of earth fell down over track in many places burying it as under a mountain... Had to stop the engine half a mile before reaching the tunnel, at a cut where a car had been buried under a slide of earth." (Strentzel 1850-82)

www.ingramcontent.com/pod-product-compliance
Lightning Source LLC
Chambersburg PA
CBHW041723210326

41598CB00007B/766